Introduction to
AUDIO ANALYSIS

Introduction to
AUDIO ANALYSIS:
A MATLAB Approach

THEODOROS GIANNAKOPOULOS

AGGELOS PIKRAKIS

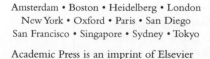

Amsterdam • Boston • Heidelberg • London
New York • Oxford • Paris • San Diego
San Francisco • Singapore • Sydney • Tokyo

Academic Press is an imprint of Elsevier

Academic Press is an imprint of Elsevier
The Boulevard, Langford Lane, Kidlington, Oxford OX5 1GB, UK
225 Wyman Street, Waltham, MA 02451, USA
525 B Street, Suite 1800, San Diego, CA 92101-4495, USA

First edition 2014

MATLAB® is a registered trademarks of The MathWorks, Inc.

For MATLAB and Simulink product information, please contact:
The MathWorks, Inc.
3 Apple Hill Drive
Natick, MA, 01760-2098 USA
Tel: 508-647-7000
Fax: 508-647-7001
E-mail: info@mathworks.com
Web: mathworks.com

British Library Cataloguing in Publication Data
A catalogue record for this book is available from the British Library

Library of Congress Cataloging-in-Publication Data
A catalog record for this book is available from the Library of Congress

ISBN: 978-0-08-099388-1

For information on all Academic Press publications
visit our web site at books.elsevier.com

Working together
to grow libraries in
developing countries

www.elsevier.com • www.bookaid.org

CONTENTS

PREFACE

This book attempts to provide a gentle introduction to the field of audio analysis using the MATLAB programming environment as the vehicle of presentation. Audio analysis is a multidisciplinary field, which requires the reader to be familiar with concepts from diverse research disciplines, including digital signal processing and machine learning. As a result, it is a great challenge to write a book that can provide sufficient coverage of the important concepts in the field of audio analysis and, at the same time, be accessible to readers who do not necessarily possess the required scientific background.

Our main goal has been to provide a standalone introduction, involving a balanced presentation of theoretical descriptions and reproducible MATLAB examples. Our philosophy is that readers with diverse scientific backgrounds can gain an understanding of the field of audio analysis, if they are provided with basic theory, in conjunction with reproducible experiments that can help them deal with the theory from a more practical perspective. In addition, this type of approach allows the reader to acquire certain technical skills that are useful in the context of developing real-world audio analysis applications. To this end, we also provide an accompanying software library which can be downloaded from the companion site and includes the MATLAB functions and related data files that have been used throughout the text.

We believe that this book is suitable for students, researchers, and professionals alike, who need to develop practical skills, along with a basic understanding of the field. The book does not assume previous knowledge of digital signal processing and machine learning concepts, as it provides introductory material for the necessary topics for both disciplines. We expect that, after reading this book, the reader will feel comfortable with various key processing stages of the audio analysis chain, including audio content creation, representation, feature extraction, classification, segmentation, sequence alignment and temporal modeling. Furthermore, we believe that the study of the presented case studies will provide further insight into the development of real-world applications.

This book is the product of several years of teaching and research and reflects our teaching philosophy, which has been shaped via our interaction with our students and colleagues, and to whom we are both grateful. We hope that the will prove useful to all readers who are making their first steps in the field of audio analysis. Although we have made an effort to eliminate errors during the writing stage, we encourage the reader to contact us with any comments and suggestions for improvement, in either the text or the accompanying software library.

Theodoros Giannakopoulos and Aggelos Pikrakis
Athens, 2013

For access to the software library and other supporting materials, please visit the companion website at: htpp://booksite.elsevier.com/9780080993881

ACKNOWLEDGMENTS

This book has improved thanks to the support of a number of colleagues, students, and friends, who have provided generous feedback and constructive comments, during the writing process. Above all, T. Giannakopoulos would like to thank his wife, Maria, and his daughter, Eleni, for always being cheerful and supportive. A. Pikrakis would like to thank his family for their patience and generous support and dedicates this book to all the teachers who have shaped his life.

LIST OF TABLES

LIST OF FIGURES

CHAPTER 1

Introduction

Contents

During recent years we have witnessed the increasing availability of audio content via numerous distribution channels both for commercial and non-profit purposes. The resulting wealth of data has inevitably highlighted the need for systems that are capable of analyzing the audio content in order to extract useful knowledge that can be consumed by users or subsequently exploited by other processing systems.

Before we proceed, it is important to note that, although in this book the term 'audio' does not exclude the speech signal, we are not focusing on traditional speech-related problems that have been studied by the research community for decades, e.g. speech recognition and coding. It is our intention to provide analysis methods that can be used to study various audio modalities and their relationships in mixed audio streams. Consider, for example, the task of segmenting a radio broadcast into homogeneous parts that contain either speech, music, or silence. The development of a solution for such a task demands that we are familiar with various audio modalities and how they affect the performance of segmentation algorithms in audio streams. In other words, we are not interested in providing solutions that are well tailored to specific audio types (e.g. the speech signal) but are not applicable to other modalities.

As with several other types of media, the automatic analysis of audio signals has been gaining increasing interest during the past decade. Depending on the storage/distribution format, the respective audio content classes, the co-existence of other media types (e.g. moving image), the user requirements, the data volume, the application context, and numerous other parameters, a diversity of applications and research trends have emerged to deal with various audio analysis tasks. The following list includes both speech and non-speech tasks so as to provide a general idea of the trends in several popular areas of speech/audio processing:

- Speech recognition: this is the task of 'translating' a speech signal to text using computational tools. Speech recognition is the oldest domain of audio analysis, but it is beyond the purpose of this book to provide a detailed study on speech recognition. We only present generic dynamic time warping and temporal modeling techniques that can also be applied on other audio signals.
- Speaker identification, verification and diarization: These speaker-related tasks focus on designing methods that discriminate between different speakers. Speaker identification and verification can be useful in the development of secure systems and speaker diarization, being able to answer the question 'who spoke when?', can be used in conversation summarization systems.
- Music information retrieval (MIR): due to the huge increase in the amount of available digital music data during the past few years, there has been an increasing need for the automatic analysis of this type of data. MIR focuses on automatically extracting information from the music signal for the purposes of content tagging, intelligent indexing; retrieval; browsing of music tracks; recommendation of new tracks based on music content (possibly combined with user preferences and collaborative knowledge); segmentation of music tracks, generation of summaries; extraction of automated music transcriptions, etc.
- Audio event detection: this is the task of detecting audio events in audio streams. There can be numerous related applications, like audio-based surveillance, violence detection, and intrusion detection, to name but a few.
- Speech emotion recognition: this is the task of predicting the speaker's emotional state (anger, sadness, etc.) using speech analysis techniques. Emotion recognition has been gaining increasing interest during the last decade. The audio stream is either used independently, or in collaboration with visual cues (e.g. facial features). Emotion recognition is expected to play an important role in the next-generation human-computer interaction systems, but it can be also be used to enhance the functionality of other systems that perform retrieval and multimedia content characterization tasks.
- Multimodal analysis of the movie content: this task aims to automatically recognize events and classes in movies based on audio, visual, and textual information. The audio cues can contain rich information regarding events like the existence of music, speech, sound effects (gunshots, human

fights), emotions, etc. The resulting metadata can serve indexing and fast browsing purposes in the context of next-generation multimedia systems.

The purpose of this book is to serve as a standalone introduction to audio signal analysis by providing a sufficient *theoretical background* for many state-of-the-art techniques, along with a large number of reproducible *MATLAB* examples. It is important to note that it is not our intention to demand that the reader be familiar with concepts from a variety of disciplines, such as signal processing and machine learning, although, of course, knowledge improves the reading experience. However, in each chapter, we focus on providing a smooth transition from introductory issues to more advanced ones, assuming that the reader is a beginner in the field. For example, we present the classification of audio segments but instead of assuming that the reader has knowledge of the respective pattern recognition concepts, we provide an introduction to the subject, ensuring that we: (a) complement the description with MATLAB examples and (b) evaluate the audio analysis domain (e.g. discuss a binary classifier via a speech-music discrimination example). Furthermore, the first chapters of the book introduce basic signal processing concepts like sampling and frequency representations.

1.1. THE MATLAB AUDIO ANALYSIS LIBRARY

Further to the necessary theoretical background, we also provide a complete set of MATLAB files that constitute the MATLAB Audio Analysis Library of this book. Where we find it useful from a pedagogical perspective, parts of the code are listed in the book. However, in most cases, the complete MATLAB code is omitted. We prefer to describe how to 'call specific functions,' to report on what to expect, to present and discuss the results, and so on.

The accompanying library is an important companion to the book that is aimed at helping the reader to understand the related theory and experiment with their own audio analysis solutions. A list of the available MATLAB functions, along with brief descriptions, is given in the Appendix of this book.

1.2. OUTLINE OF CHAPTERS

Chapter 2 provides information and techniques for the basic issues related to the creation, representation, playback, recording, and storing of audio signals in MATLAB. Although the focus of the chapter is on practical issues, we also describe the basic theory of content creation. At the end of the chapter, we describe the process of breaking an audio signal into short-term windows to enable audio analysis on a short-term basis. This is in preparation for the next two chapters, as frequency representations and feature extraction both require the short-term processing stage of the signal.

In Chapter 3 we present methods for representing audio signals in the frequency domain, mostly focusing on the discrete Fourier transform. In addition, we provide a basic description of filtering techniques by Means of MATLAB examples.

Chapter 4 presents a wide range of features from the time and frequency domains, that have been widely used in various audio analysis approaches. Sufficient theoretical background is provided for each feature along with MATLAB code from the companion Library. In addition, the discrimination ability of each feature for particular types of sound is demonstrated.

After the reader has been introduced to a series of audio features and their discrimination capabilities for selected audio classes, Chapter 5 describes the task of classifying unknown audio segments of homogeneous content. For instance, in Chapter 4, the reader will learn that, in many cases, the standard deviation of short-time energy can discriminate between music and speech segments, whereas in Chapter 5 the reader will learn how to combine several feature statistics in the context of a classification procedure. To this end, we provide necessary theoretical background for a series of standard classification techniques, including support vector machines, decision trees, and the k-nearest-neighbor method. The reader will be also introduced to generic performance measures and validation methods for the estimation of the performance of a classifier. The chapter concludes with the presentation of performance measurements for a series of typical audio classification tasks (e.g. music vs speech classification).

Chapter 6 presents another processing stage of vital importance in audio analysis: the segmentation stage. The goal of this task is to split an uninterrupted audio signal into segments of homogeneous content. In this chapter we focus on two general segmentation methodologies: those that exploit prior knowledge of the audio types involved (and therefore embed some type of classification mechanism) and those that are unsupervised (or semi-supervised). We also study segmentation tasks of particular interest, e.g. silence removal and speaker diarization.

Chapter 7 also looks at classification techniques, but from a different point of view to Chapter 5. The focus with this chapter is on template matching and hidden Markov modeling and its goal is to exploit the temporal evolution of the feature sequence, whereas the methods in Chapter 5 are based on statistical averages of the feature sequences on a mid-term and long-term basis.

Finally, Chapter 8 presents a series of music information retrieval applications, including music thumbnailing; tempo and music meter induction; and music content visualization. It presents methods that combine audio analysis techniques from earlier chapters, in the context of music information retrieval applications. However, in some cases new concepts are introduced.

The book also provides an extensive appendix that covers the following:
- The MATLAB functions and data of the software library.
- A short description of libraries and software packages available on the Web that are related to audio analysis and pattern recognition. This is not limited to MATLAB approaches, as Python and C/C++ packages are also included.
- Provides a list of datasets that are available on the Web, that can be used as training and evaluation data for several audio analysis tasks.

1.3. A NOTE ON EXERCISES

At the end of each chapter, we provide a set of exercises. The purpose of the assignments is to cover the content of the respective chapter, while triggering the interest of the reader to extend his/her knowledge on audio analysis tasks. The exercises have been graded based on their difficulty with a five-level grading system, as presented in Table 1.1.

Table 1.1 Difficulty Levels of the Exercises

Level	Description
D1	Simple questions that require either trivial MATLAB coding or a short answer.
D2	Exercises that test the reader's understanding of the MATLAB code of the respective chapter.
D3	More complex assignments that require either a combination of knowledge presented in the chapter or more advanced critical thinking. May require writing MATLAB code.
D4	Larger assignments that require more MATLAB coding. Can be used as weekly or monthly projects in the context of an audio analysis course.
D5	The most challenging level includes exercises that require extensive MATLAB coding, external bibliographic references, or external MATLAB libraries. Exercises of this level can be used as final projects.

We encourage readers who are new to the field to read all chapters from the beginning of the book and complete all the exercises. More advanced readers can skip certain parts of the book based on their experience. The book is suitable for undergraduate courses on audio, especially when a diverse audience is involved. The MATLAB code can be also used to set up laboratory exercises and projects.

Getting Familiar with Audio Signals

Contents

The goal of this chapter is to provide the basic knowledge and techniques that are essential to create, playback, load, and store the audio signals that will be analyzed. Therefore, we focus on certain practical skills that you will need to develop in order to prepare the audio signals on which the audio analysis methods of later chapters will be applied. Although we have intentionally avoided overwhelming the reader with theory at this stage, we provide certain pointers to the most important theoretical issues that will be encountered. Our vehicle of presentation will be the MATLAB programming environment and even though the source code may not be directly reusable in other software development platforms, the presented techniques are commonplace and highlight certain key issues that emerge when it comes to working with audio signals, e.g. the difficulties in manipulating voluminous data.

2.1. SAMPLING

Before we begin, it is worth noting that, in principle, digital audio signals can be easily represented in the MATLAB programming environment by means of vectors of real numbers, as is also the case with any discrete-time signal. The term *discrete time* refers to the fact that although in nature time runs on a continuum, in the digital world we can only manipulate *samples* of the real-world signal that have been drawn on discrete-time instances. This process is known as *sampling* and it is the first stage in the creation of a digital signal from its real-world counterpart. The second stage is the quantization of the samples, but for the moment we will ignore it.

To give an example, the first sample may have been taken at time 0 (the time when measurement commenced), the second sample at 0.001 s, the third one at 0.002 s, and so on. In this case, the time instances are equidistant and if we compute the difference between any two consecutive time instances the result is $T_s = 0.001$ s, where T_s is known as the sampling period. In other words, one sample is drawn every T_s seconds. The inverse of T_s is the celebrated sampling frequency, i.e. $F_s = \frac{1}{T_s}$ Hz. The sampling frequency is measured in Hz and in this example it is equal to $Fs = \frac{1}{0.001} = 1000$ Hz i.e. 1000 samples of the real-world signal are taken every second.

A major issue in the context of sampling is how high the sampling frequency should be (or equivalently, how short the sampling period has to be). It turns out that in order to successfully sample a continuous–time signal, the sampling frequency has to be set equal to at least twice the signal's maximum frequency [1]. This lower bound of the sampling frequency is known as the Nyquist rate and it ensures that an undesirable phenomenon known as *aliasing* is avoided. More details on aliasing will be given later in this book but for the moment it suffices to say aliasing introduces distortion, i.e. it reduces the resulting audio quality.

2.1.1. A Synthetic Sound

In order to demonstrate the sampling process, we could simply start with a recording device and set its recording parameters to appropriate values so as to achieve a good quality recorded signal. However, we embark on an alternative route; we start with an example that creates a synthetic sound. In other words, we are making a first attempt to simulate the sampling stage of the recording procedure by employing our basic sampling knowledge. More specifically, in the following script, three sinusoidal signals (tones) of different frequencies are synthetically generated and their sum is computed and plotted. The resulting signal is, therefore, a sum of tones and the tone of maximum frequency (900 Hz) determines the Nyquist rate. Therefore, in this case, a sampling frequency of at least 1800 Hz is required. The comments on each line explain the respective command(s). This is a style of presentation that we have preserved throughout this book.

```
% script1.m
% Demonstrates the generation of synthetic tones

Fs = 16000; Ts = 1/Fs;       % Sampling frequency, sampling period
time = 0:Ts:0.1;             % Define when sampling occurs
                             % (sampling lasts for 0.1 secs):
Freqs = [250 550 900];       % These are the frequencies of the signals
Xs = zeros(length(Freqs), length(time));

for i=1:length(Freqs)        % Create one audio signal (tone) per frequency
    Xs(i,:) = cos(2*pi*Freqs(i)*time);
end

x = sum(Xs);                 % The final sum of tones
x = x ./ max(abs(x));        % Normalize x

figure; plot(time, x); axis([0 time(end) -1 1]); % Plot x:
xlabel('Time (sec)'); ylabel('Signal Ampl.'),
title('A simple audio signal')
```

In order to understand how the script works, consider a simple tone (sinusoidal signal) in continuous time, with frequency 250 Hz, i.e. $x_a(t) = A\cos(2\pi 250t)$, where A is the amplitude of the tone (phase information has been ignored for the sake of simplicity). In our case, $A = 1$ for each tone, meaning that each tone's maxima and minima occur at $+1$ and -1, respectively. If T_s is the sampling period then the samples are taken at multiples of Ts, nTs, where n is a positive integer and $nT_s <= 0.1$. In other words, our example first needs to implement the equation $x_a(nT_s) = \cos(2\pi fnT_s)$ for each tone and then sum the resulting signals. Note that the final audio signal is normalized by dividing it with its maximum value and as a result, the values in vector x lie in the range $[-1, 1]$. Figure 2.1 presents the audio signal that is generated by the above script.

2.2. PLAYBACK

A useful MATLAB command, which sends a vector (stream) of samples, x, to the sound card device for playback purposes is the `sound(x,fs,nbits)` command, where `fs` is the sampling frequency based on which x will be reproduced and `nbits` is the number of bits that are used to represent each sample (this number is also known as *the sample's depth* and is determined

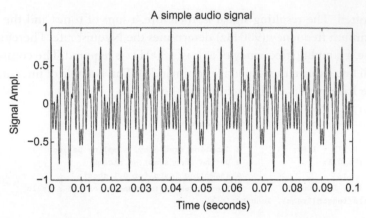

Figure 2.1 A synthetic audio signal.

during the quantization of samples). If the sampling frequency is not provided then a default value is assumed by MATLAB (around 8 KHz). Similarly, depending on the operating environment, the default bit representation may utilize 8 or 16 bits. In order to avoid distortion during the playback operation, the values in x have to be in the range $[-1,1]$. If you want to listen to the signal that was generated by the example in Section 2.1.1, type:

```
sound(x,Fs);
```

It is important to note that when using MATLAB with operating systems other than the MS Windows environment, certain sound playback issues may be encountered. For example, in the Ubuntu OS, the sound() command may lead to an error due to sound card compatibility issues. In order to circumvent such situations, we can resort to an alternative way to play a sound: we first store it in a temporary audio (WAVE) file and then we 'play' the file using an appropriate system utility. For example, in the Ubuntu OS, this functionality can be achieved by the aplay command-line utility, which automatically determines the sampling rate, number of bits per sample, and other properties of the sound-file format and then plays the respective sound. In this line of thinking, the following function can be used to reproduce a sound both in the MS Windows and Ubuntu operating systems:

```
function soundOS(x, fs, nBits)

% function soundOS(x, fs, nBits)
%
% This function is an alternative to the Matlab sound() function for
% playing sounds in both the WINDOWS and UBUNTU operating systems
```

```
if nargin==1
    fs = 8000; % define default sampling rate
    nBits = 16; % define default value for bits/sample
end
if nargin==2    % if fs has been provided by the user
    nBits = 16; % define default value for bits/sample
end

if isunix
    % generate a random file name for a temporary .wav file
    wavFileNameTemp = sprintf('temp_sound_%d.wav', round(rand(1)*100));
    % store the sound in the temporary wav file
    wavwrite(x, fs, nBits, wavFileNameTemp);
    % playback the file using the aplay ubuntu command
    system(['aplay ' wavFileNameTemp]);
    % delete the temporary file:
    system(['rm ' wavFileNameTemp]);
else
    % otherwise use the default sound() command:
    sound(x, fs, nBits);
end
```

Notes:
- The `system()` MATLAB function employs the operating system to execute the command which is given to its input as a string argument.
- The `isunix` MATLAB returns 1 for Unix versions of MATLAB and 0 otherwise.
- The `wavwrite()` function is used for storing an audio signal to a WAVE file and will be further described in Section 2.4.

2.3. MONO AND STEREO AUDIO SIGNALS

In MATLAB, a column vector represents a single-channel (monophonic—MONO) audio signal. Similarly, a matrix with two columns refers to a two-channel signal (stereophonic—STEREO), where the first column represents the left channel and the second column represents the right channel. The following code creates a STEREO signal. The left channel contains a 250 Hz tone (cosine signal) and the right channel a 450 Hz tone. Figure 2.2 provides a separate plot of each channel over time.

```
% script2.m
% Generation, representation and playback of STEREO signals in Matlab

Fs = 16000; Ts = 1/Fs;        % Sampling freq, period
time = 0:Ts:0.1;              % Define time vector (0.1 secs duration)
```

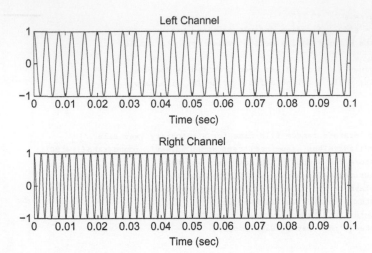

Figure 2.2 A STEREO audio signal.

```
% Signal computation:
xLeft  = cos(2*pi*250*time)';    % Left channel
xRight = cos(2*pi*450*time)';    % Right channel
x = [xLeft xRight];              % Compose STEREO signal

% Plot x:
figure;
subplot(2,1,1); plot(time, x(:,1)); xlabel('Time (sec)'); title('Left Channel');
subplot(2,1,2); plot(time, x(:,2)); xlabel('Time (sec)'); title('Right Channel');
sound(x,Fs);                     % Play x:
```

> *Note:* When x has 2 columns, i.e. when the corresponding signal is in STEREO mode, we usually want to covert it to a monophonic version before we actually start processing it. A simple way to achieve this conversion is by averaging the two signal channels: $x = \text{sum}(x,2)$;

2.4. READING AND WRITING AUDIO FILES

2.4.1. WAVE Files

A common file format for uncompressed audio is the WAVE format which is built on pulse-code modulation (PCM) technique for the quantization of samples. The associated files are expected to have a '.wav' extension. In MATLAB, the audio content of such files can be easily loaded into the workspace using the `wavread()` function (type help wavread for a detailed description of this function). The most important input argument of this function is a string which specifies the name of the WAVE file to be read:

```
[x,Fs,nBits] = wavread('test');
```

Wavread() returns the audio signal, x; the sampling frequency, Fs, with which the signal was recorded (or synthetically generated); and the number of bits per sample, nBits. If the file contains a MONO audio signal, then x is a column vector, whereas, if the file contains a STEREO sound then x is a matrix of two columns, one column per channel. Wavread() can also be used to read excerpts (segments) of the audio content. The following call to `wavread()` loads into the workspace just the first N samples of the file:

```
[x,Fs,nBits] = wavread('test',N);
```

It is also possible to read samples $N1$ through $N2$ from each channel by using the following syntax:

```
[x,Fs,nBits] = wavread('test',[N1 N2]);
```

Note: If an audio file is too large (e.g. due to a lengthy recording and/or high sampling rate), memory allocation problems may arise when trying to read all the content of the file with a single call to `wavread()`. In such cases, it is desirable to read the audio content progressively (in blocks), using the aforementioned alternative syntax. For more details, the reader may refer to Section 2.5.

It is also possible to simply retrieve the length of the audio stream without actually reading the signal samples. This is feasible due to the structure of WAVE files. This structure has been standardized in the respective format specification and is reflected on the "header" of the file. The header is a chunk of data that reside in the beginning of the file and give various details including sampling frequency, depth of samples, number of channels, order with which the samples are stored, and so on. The following call to wavread() will therefore only retrieve header information from file 'test.wav.'

```
[SIZE, Fs, nBits] = wavread('test', 'size');
```

The first output variable, SIZE, contains two elements: the number of samples and the number of channels. As a result, to compute the time duration of the audio signal (in seconds), type

```
SIZE(1)/Fs
```

In order to **store** an audio signal, x, into a WAVE file, the `wavwrite()` function can be used (type help wavwrite for a detailed description of input and output arguments). In summary, the input arguments for this function are: the signal, x (column vector or two-column matrix); the sampling frequency, Fs; the number of bits per sample, and the name of the file to create.

```
wavwrite(x,Fs,16,'test')
```

Note that the above function call results in a destructive operation, in the sense that, if the file already exists, then its content will be overwritten. For the reader who is interested in the technicalities of wavwrite(), we provide a few more details: if x is a vector of floating-point numbers, then the number of bits per sample can be 8, 16, 24, or 32 and each element of x should preferably lie in the range $[-1, 1]$. If a value falls outside the previous range, an operation known as 'clipping' will be performed (one of the exercises in this chapter involves the clipping technique). The elements of x may also be of other data types, e.g. of the int16 data type (16-bit integers covering the range of integers $-32768, \ldots, +32767$). In such cases, the reader is encouraged to consult the help utility of the function in order to determine which values are allowable for the number of bits per sample.

Note: The `wavinfo()` MATLAB function can be used to extract the header of a WAVE file and return it as a MATLAB structure.

2.4.2. Manipulating Other Audio Formats

MATLAB supports some other standard audio formats further to the WAVE format. For example, function `auread()` can be used to load audio stored in '.au' files. The *Au file format* [2] provides a simple interface for audio storage and its philosophy bears certain similarities with the WAVE format specification. The *Au file format* has a simple header and uses simple encoding schemes for the audio data, including 8-bit μ-law and 8-,16-,24-,32-bit linear PCM. As with the syntax of `wavread()`, `auread()` also returns the signal samples in an array and the sampling frequency and bits per sample in separate output variables.

Another very popular format, for which it is useful to have a MATLAB encoder and decoder, is the MP3 format, which is based on the principles of perceptual coding [3, 4]. MP3 is a lossy format, in the sense that the decoded signal is not identical to the original one. Until recently, MATLAB did not support I/O for MP3 files and developers had to rely on 3d-party software. Lately, the `audioread` function was added to the MATLAB environment, making it possible to manipulate MP3 audio data from the workspace. Note that `audioread()` is not restricted to MP3 files and serves as a root function which is aimed at addressing various audio data formats. Its syntax is similar to the syntax of `wavread()` and `auread()`.

An alternative way to read audio data of (almost) every popular format, including the aforementioned formats, is by using the *ffmpeg* command-line tool of the FFmpeg multimedia framework ([1]). At the time of writing, the FFmpeg suite of tools was a GNU-LGPL and GNU-GPL licensed cross-platform framework. The goal of the FFmpeg project is to allow users to engage with various multimedia processing operations, including the loading and storage of audio data and the conversion among audio data types. The FFmpeg framework uses its own codecs (implementations of coding-decoding schemes) for data processing and its functionality is available to program developers (e.g. C/C++ program developers) by means of an appropriate application programming interface (API). In the rest of this section, we present on a by-example basis, the functionality of the FFmpeg tool of the FFmpeg project with respect to audio processing; and we show how this functionality can be made available to MATLAB routines (the toolkit of the FFmpeg project includes the *ffserver, ffplay*, and *ffprobe* tools).

For the moment, let us ignore MATLAB and assume that we want to transform an MP3 file, say 'file.mp3' to the WAVE file 'output.wav.' This type of processing is also known as *transcoding* because we transform a data stream from one coding scheme to another. To this end, start a terminal window and type:

```
ffmpeg -i file.mp3 output.wav
```

It is also possible to transcode selected parts of the audio stream. For example, if you type:

```
ffmpeg -i file.mp3 -t 5 output.wav
```

then the first 5 s of the audio signal in file.mp3 are transcoded and stored to file output.wav.

Another useful parameter that can be passed to the FFmpeg tool as a command-line argument is the `-ar <value>` flag, which sets the sampling

---

[1] http://www.ffmpeg.org/.

rate for the output file. In a similar manner, flag -ac sets the number of output channels. Therefore, the command

```
ffmpeg -i file.mp3 -ar 16000 -ac 1 output.wav
```

extracts the audio signal from *file.mp3*, resamples it at 16 KHz, converts it to a single-channel representation, and stores it to file *output.wav*.

In the library of MATLAB functions that we provide with this book, you will find the function mp3toWav(), which simply uses MATLAB function *system()* to call the FFmpeg tool in order to transcode an MP3 file to the WAVE format. In addition, function mp3toWavDIR() transcodes all MP3 files in a folder to their WAVE counterparts. For each successful audio stream conversion, a new file is created, preserving the original filename and modifying the extension to address the new file format.

Note:

If you have MATLAB installed on a Linux distribution, you may encounter certain problems when calling external applications using the system() function, so mp3toWav() might not work. In particular, when calling the FFmpeg tool, the following error message may emerge:

```
ffmpeg: MATLAB_PATH/bin/glnxa64/libstdc++.so.6:
version 'GLIBCXX_3.4.15' not found (required by
/usr/lib/x86_64-linux-gnu/libjack.so.0)
```

You can remedy this situation by updating the respective link. To do this, type in a terminal window the following commands:

```
sudo updatedb
locate libstdc++.so.6
sudo ln -sf /usr/lib/i386-linux-gnu/libstdc++.
so.6.0.17 /MATLAB_PATH/bin/glnxa64/libstdc++.
so.6
```

Of course, the path names may vary depending on your configuration. For example, the path of the link could be /MATLAB_PATH/sys/os/glnxa64/libstdc++.so.6 and not the one above. In any case, the error message which is generated by calling the FFmpeg tool should be able to highlight the respective paths.

Finally, note that you will encounter several examples of MATLAB code on the Internet where the FFmpeg toolkit is used to handle various types of multimedia content. The MATLAB Central website for File Exchange can be a good source for the interested reader. An example is the `mmread()` function (authored by Micah Richert, available via the Mathworks File Exchange website,[2]) which is based on the joint use of FFmpeg and AVbin.

2.5. READING AUDIO FILES IN BLOCKS

As explained in Section 2.5, if an audio file is very large, loading all its contents into memory may be impractical. Consider, for example, a .wav file in which a 3-minute music track has been stored. The sampling frequency is assumed to be 44100 Hz (CD quality recording) and two channels have been used (STEREO mode). If we consider that MATLAB uses by default double-precision floating-point arithmetic for its variables (i.e. each variable is by default 8 bytes long), then the total number of bytes needed to load this music track into memory is: $3 \cdot 60 \cdot 44100 \cdot 2 \cdot 8 =\simeq 200$ MB. Usually, there is no need to keep the whole signal in memory all the time, because, in most applications, the audio signals are processed on a short-term basis. (See Section 2.7 for more details on short-term processing.)

The goal of the next example is to demonstrate block-based audio processing, a simple technique to overcome the need for loading and keeping all the contents of the audio file into memory. The corresponding function is `readWavFile()`. You will need to execute the following steps:

- Determine the total number of audio samples and the sampling frequency, using the 'size' option in the call to the `wavread()` function:

```
[a, fs] = wavread(fileName, 'size');
numOfSamples = a(1);
nChannels = a(2);
curSample = 1;
```

- Use a while loop for reading one block of the audio signal each time. The limits of each block (indices of leftmost and rightmost sample) need to be provided in the call to the `wavread()` function:

```
while (curSample ≤ numOfSamples)
    N1 = curSample;
    N2 = curSample + BLOCK_SIZE − 1;
    if (N2>numOfSamples)
        N2 = numOfSamples;
    end
    tempX = wavread(fileName, [N1, N2]);
    curSample = curSample + BLOCK_SIZE;
end
```

Concerning response times, the block-reading mode introduces a small delay due to the repeated calls to the `wavread()` function (one call per

---

[2] http://www.mathworks.com/matlabcentral/fileexchange/8028-mmread.

Table 2.1 Execution Times for Different Loading Techniques

Reading Mode	6 min	48 min
No blocks (simple `wavread()`)	1.72	2.40
Block = 0.5 min	1.83	2.80
Block = 2 min	1.82	2.65
Block = 5 min	1.78	2.53

block). However, this is not a crucial issue as it can be also seen in Table 2.1, where the execution times are presented for different files on a standard laptop. Note that if your application demands for efficient IO handling, then you probably need to resort to an alternative implementation of the `wavread()` function. For example, you can re-engineer the `wavread()` function so that the audio file is left open throughout the whole reading operation, thus reducing certain IO overheads.

2.6. RECORDING AUDIO DATA

2.6.1. Audio Recording Using the Data Acquisition Toolbox

In order to read data from a sound card, you can use the Data Acquisition Toolbox of MATLAB. In addition to sound recording, this toolbox also provides functionality for controlling various types of data acquisition hardware. In order to use the Data Acquisition Toolbox for audio recording, you need to follow the steps given in Table 2.2.

Table 2.2 Sound Recording Using the Data Acquisition Toolbox

Step	Description
1	Create a device object
2	Add channels
3	Configure recording properties
4	Initialize recording
5	Trigger recording
6	Acquire data
7	Terminate process

It is worth noting that if steps 4–6 are nested inside a loop, then each data block can be processed separately, right after it has been acquired. This is useful when it is desirable to perform online audio processing operations.

Let us now review the basic steps of the recording procedure in more detail. The first two steps create a sound card object and add an audio channel to that object. The audio object serves as an operation handler, i.e. the object is the gateway via which the recording procedure is actually implemented:

```
ai = analoginput('winsound');
addchannel(ai,1);
```

The winsound string, which is given as input to the *analoginput* function defines that the sound card object is created using an MS Windows driver. Our next concern is to set values for a number of recording parameters, namely, the sampling frequency (*SampleRate*), the length of the recording block (*SamplesPerTrigger*), and the triggering mode (*TriggerType*). The last parameter defines that the recording times are manually controlled. As can be seen in the following code, for each recording parameter we need to set the value of the respective object attribute:

```
Fs = 16000;
duration = 1.0;
nBlocks = 20;
set (ai, 'SampleRate', Fs);
set(ai, 'SamplesPerTrigger', duration*Fs);
set(ai,'TriggerType','Manual');
```

Then, inside a loop (i.e. for each block), we initialize (start) and begin (trigger) the recording process:

```
start(ai);
trigger(ai);
```

Inside the loop, we use a variable (*x*) to store each block of acquired data:

```
x = getdata(ai, duration * Fs);
```

After the data acquisition loop has ended, we clear the previously created acquisition handler:

```
delete(ai);
clear ai;
```

The reader may refer to `script3.m` for the complete version of the above audio recording procedure. The functionality in `script3.m` can be summarized as follows:

- Twenty blocks of audio data are recorded in total. Each block is 1.0 s long.

- A buffer is used to store the complete audio recording. This is done mainly for educational purposes and it may not be necessary in many applications, for reasons that were previously stated.
- At each iteration, while the current block is being recorded, the intensity of the signal (in dB) is computed for the previous block and is plotted using a simple bar plot.
- In the end, the contents of the audio buffer are stored in a WAVE file.

> *Note:* Operating system-related issues may emerge when using the `analoginput()` and related functions, so the reader is encouraged to refer to the respective technical descriptions for assistance.

2.6.2. Audio Recording Using the Audio Recorder Function

The recording procedure of Section 2.6.1 is just one technique for sound recording with MATLAB. An alternative approach is to use the `audiorecorder` function. Note that this methodology does not support disk buffering. Instead, it stores the audio samples in the computer memory. It is, therefore, not the best solution when the expected recording duration is very long. If, however, the function is used in a block-processing context, as in the following example, the memory usage issues can be circumvented because intermediate audio data blocks can be stored on a disk.

In order to use the `audiorecorder()` function, we first need to set the parameters of the recording procedure (i.e. the sampling frequency, the number of bits per sample, and the number of channels). We then use the `audiorecorder` constructor to create a recording object (`recObj1`):

```
Fs = 16000;
nBits = 16;
nChannels = 1;
recObj1 = audiorecorder(Fs, nBits, nChannels);
```

After the recording object has been created, the actual recording process is started:

```
record(recObj1, BlockSize);
```

`BlockSize` is the duration of each block to be recorded and `record` is a non-blocking function, i.e. a function that returns immediately, making it suitable for online (block-based) processing applications. In the context of block-processing procedures, a MATLAB timer can be adopted for synchronization purposes (e.g. using MATLAB's `cputime()` function). In that case, a loop is used and at each iteration, we measure the elapsed time since the beginning of the recording of the current block. If the elapsed time exceeds the desired length (in seconds) of the recording block, then we: (a) stop the recording process, (b) get the audio data of the last block and store it in a variable, and (c) restart the recording process, along with the MATLAB timer:

```
T1 = cputime;
while (1)
  T2 = cputime;
  if T2-T1>BlockSize
    stop(recObj1);              % stop recording data
    x = getaudiodata(recObj1);  % get the current block data
    T1 = cputime;
    record(recObj1, BlockSize); % start recording buffer
    ....
```

The process is repeated until the desired number of blocks has been acquired. The `script4` m-file provides a detailed demonstration of the `audiorecorder()` function in the context of the block-processing (online) mode of operation (the design of the rest of the script has been kept similar to the example of Section 2.6.1).

Note: Instead of using the `record()` function, we can alternatively use the `recordblocking()` function, which does not return control until recording completes. The second argument for that function is the duration of the data to be recorded. In this way, the resulting code becomes rather simple:

```
% initialize the audiorecorder object:
recObj1 = audiorecorder(16000, 16, 1);
% start the recording (blocking mode for 5 seconds):
recordblocking(recObj1, 5);
% get the data:
x = getaudiodata(recObj1);
```

The `script4` m-file is using 'manually' controlled synchronization. However, MATLAB provides yet another method for the online processing of recorded data, i.e. the mechanism of *callback functions*. More specifically, a property named `TimerFcn` of the `audiorecorder` object can be set to define the callback function that will be executed repeatedly during the recording procedure. To specify the periodicity of repetitions, it is necessary to set the `TimerPeriod` property of the `audiorecorder` object. For example:

```
recObject = audiorecorder(16000, 16, 1);
set( recObject, 'TimerFcn', @someFunction, 'TimerPeriod', 0.5);
```

In the above lines of code, `someFunction` is the name of the callback function to be executed every 0.5 s. An example of such a callback function is the `audioRecorderTimerCallback()` function that we provide in the software library of this book. That function is used by `audioRecorder Online`, which simply initializes the `audiorecorder` object. In other words, `audioRecorderTimerCallback()` defines the processing steps which need to be applied on the recorded data at every iteration. In the provided example, we simply plot the signal samples at each recording stage. In order to execute this callback-based audio recording procedure, simply call the initialization function from the MATLAB workspace:

```
audioRecorderOnline
```

2.7. SHORT-TERM AUDIO PROCESSING

In most applications, the audio signal is analyzed by means of a so-called *short-term* (or *short-time*) processing technique, according to which the audio signal is broken into possibly overlapping short-term windows (frames) and the analysis is carried out on a frame basis. The main reason why this windowing technique is usually adopted is that the audio signals are non-stationary by nature, i.e. their properties vary (usually rapidly) over time [5]. Let us ignore for the moment the mathematical formalism and try to understand, by means of a simplified example, what non-stationarity stands for. More specifically, consider an audio recording consisting of a short conversation (1 s long) between two individuals, that is followed by the shout of a third person (also 1 s long). Therefore, this audio recording consists of two main events: the conversation (normal intensity signal) and the shout (high-intensity signal). It is obvious that the signal changes abruptly from the state of the conversation to the state of the shout. From a very simplified perspective, this can be considered as a change of stationarity, i.e. the properties of the signal shift from one state to another. In such situations, it would not really make sense to compute, for example, the average intensity of the samples of the whole recording because the resulting value would be dominated by the more intense samples that were recorded during the shout of the third person. Instead, it would be more useful to break the recording into short segments and compute one value of (average) intensity per segment. This is also the main idea behind short-term processing. Of course, it has to be noted that in our example, the change of stationarity was observed at the more abstract level of audio events, whereas short-term processing usually works at the microcosm of the samples of the signals.

To continue, let $x(n)$, $n = 0, \ldots, N - 1$, be an audio signal, N samples long. During short-term processing, we focus each time on a small part (frame) of the signal. In other words, we are following a windowing procedure: at each processing step, we multiply the audio signal with a shifted version of a finite duration window function, $w(n)$, i.e. a discrete-time signal which is zero outside a finite duration interval. The resulting signal, $x_i(n)$, at the ith processing step is given by the equation:

$$x_i(n) = x(n)w(n - m_i), \quad i = 0, \ldots, K - 1, \tag{2.1}$$

where K is the number of frames and m_i is the shift lag, i.e. the number of samples by which the window is shifted in order to yield the ith frame. Equation (2.1) implies that $x_i(n)$ is zero everywhere, except in the region of samples with indices $m_i, \ldots, m_i + W_L - 1$, where W_L is the length of the moving window (in samples). The value of m_i depends on the hop size (step), W_S, of the window. For example, if the window is shifted by 10 ms at each step and the sampling frequency, F_s, is 16 kHz, then, $m_i = i \cdot W_S \cdot F_s = i \cdot 0.01 \cdot 16000 = i \cdot 160$ samples, $i = 0, \ldots, K-1$. Furthermore, if $W_L = 300$ samples, then the 5th frame ($i = 4$) starts at sample index $160 \cdot 4 = 640$ and ends at sample index $160 \cdot 4 + 300 - 1 = 939$.

The above highlights the fact that the important parameters of the short-term processing technique are the length of the moving window, W_L, its hop size (step), W_S, and its type (the function used to implement the window). Usually, W_L varies from 10 ms to 50 ms. On the other hand, the window step, W_S, controls the degree of overlap between successive frames. If, for example, a 75% overlap is desired and the window length is 40 ms, then the window step has to be 10 ms. It can be easily derived that the total number, K, of short-term windows is equal to:

$$\lfloor \frac{N - W_L}{W_S} \rfloor + 1,$$

where W_L, W_S, and N were defined above and $\lfloor \ \rfloor$ is the floor operator. Figure 2.3 presents an example of short-time processing where the length of the moving window is 200 samples and its hop size is 100 samples (50% overlap between successive windows).

Concerning the window types that can be used, the simplest choice is the rectangular window (MATLAB `rectwin()` function), in which the signal is simply truncated outside the limits of the window and is left unaltered

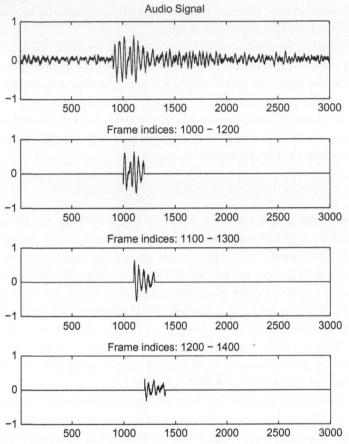

Figure 2.3 Short-term processing of an audio signal. Three consecutive frames are shown. Each frame is 200 samples long and a 50% overlap exists between successive frames.

inside the window. Equivalently, each sample inside the region defined by the window is multiplied by 1, whereas all the other samples are multiplied by 0. This is also described in the following equation:

$$w(n) = \begin{cases} 1, & 0 \leq n \leq W_L - 1, \\ 0, & \text{elsewhere.} \end{cases} \tag{2.2}$$

Further to the rectangular window, other choices (which are also supported by MATLAB) are the popular Hamming window, the Bartlett window, and the Hanning window, to name but a few. What differs between those window types is the function that determines the shape of the window, i.e. the attenuation near the edges of the window, the smoothness of the respective curve, and so on.

The stpFile() function demonstrates the short-term processing technique. The source code of this function has been kept simple for educational purposes and it is not optimal in any sense. If we go through this m-file, we can see that it first copies each frame in a temporary vector and it then multiplies the contents of the vector by the Hamming window. Of course, the Hamming window can be substituted by any valid window function. Finally, for each frame, the original and the windowed versions are plotted. From a technical perspective, if the audio file is too large then we may need to read it in blocks (as was shown in Section 2.5). In such cases, the short-term processing technique has to be applied separately on each block.

```
function stpFile(wavFile, windowLength, step)
%
% stpFile.m
%
% This function demonstrates the short—term processing of an audio signal
%
% ARGUMENTS:
% — wavFile: the name of the WAV file to be processed
% — windowLength: the length of the window (in seconds)
% — step: the window step (in seconds)
%

[x,fs] = wavread(wavFile);                      % read the WAV file

% convert window and step from seconds to samples:
windowLength = round(windowLength * fs); step = round(step * fs);
curPos = 1; L = length(x);
% compute the total number of frames:
numOfFrames = floor((L—windowLength)/step) + 1;
figure;
for (i=1:numOfFrames) % for each frame
    frame  = x(curPos:curPos+windowLength—1);   % get current frame:
    % multiply the frame with the hamming window:
    frameW = frame .* window(@hamming, length(frame));
    subplot(2,1,1); plot(frame); title('Current frame (original)');
    axis([0 length(frameW) —1 1])

    % plot windowed frame:
    subplot(2,1,2); plot(frameW); title('Current frame (windowed)');
    axis([0 length(frameW) —1 1]); drawnow;  pause(0.1);
    % DO SOMETHIN WITH 'frameW' HERE:
    % ...
    % ...
    curPos = curPos + step;
end
```

2.8. EXERCISES

1. (D1) Write a MATLAB function that reads the header of a WAVE file (the path to the file is the first argument of the function) and checks if the stored audio signal is in STEREO mode. If this is the case, then the signal is loaded, it is converted to MONO mode, and it is stored to a new WAVE file (second input argument). During this procedure the sampling frequency is left unaltered.

2. (D2) The `stpFile()` function described in Section 2.5, which can be found in the repository that accompanies the book, demonstrates a short-term processing technique that plots the short-term frames of a given signal. In this exercise, you are asked to modify the plotting procedure, so that the horizontal axis presents the time in seconds instead of showing sample indices.

3. (D3) Write a MATLAB function that:

    (a) Uses the `audiorecorder` functionality (Section 2.6.2) to record 1 s of audio data. The process is repeated twice, i.e. two 1-sec audio segments are sequentially recorded.

    (b) Creates a STEREO signal by placing each recorded segment in a separate channel.

    (c) Plays the STEREO signal.

4. (D3) If the `audiorecorder` functionality is nested inside a loop (Section 2.6.2), it can be used to record large amounts of data provided that the data are saved on a hard disk, progressively. In that way, there is no need to store the entire signal into memory. Write a MATLAB program that:

    (a) Uses the `audiorecorder` functionality to record a number of audio segments, where each segment is 10 s long. This can be achieved, either by adopting the loop-based procedure of `script4` or by using the callback-based approach (Section 2.6.2).

    (b) Stores each segment in a separate WAVE file (for example, segment 1 is stored in file output0001.wav, segment 2 in output0002.wav, and so on).

    Note that, since only 10 s of audio are stored in memory at each recording stage (and the rest on hard disk), the above procedure can be used to record arbitrarily long audio streams.

5. (D4) In this exercise you are asked to create a MATLAB function that:

    (a) Repeats the first step of the previous exercise.

    (b) For the current segment, the intensity of the signal is computed (in dB) as in `script4`, i.e. using the command `10*log10` (mean(x.^2)), where x is the vector of the samples of the segment.

(c) If the intensity level of a segment is higher than a predefined threshold (e.g. −20 dB) then the respective segment is stored in a WAVE file, otherwise the process continues with the next segment. In order to name each WAVE file that is automatically created, you may either use a timestamp that is derived from the elapsed execution time or a simple counter that counts the number of segments that survived the threshold. You may find it useful to employ MATLAB's cputime and clock functions.

Similar to the previous exercise, the computer's memory does not need to store the whole signal. However, unlike the previous exercise, not every segment is stored in a WAVE file; only segments with sufficient energy are preserved. This type of functionality can be used to monitor long periods of audio activity while storing only selected audio segments on the disk. Consider for example, an audio surveillance application, where the monitoring operation should never cease. If the solution to this exercise was adopted, there would be no need to store every audio sample on the disk, only those parts of the audio stream that correspond to high-intensity signals.

6. (D2) Section 2.4.2 described how to transcode MP3 files to the WAVE format by calling the external FFmpeg tool. The object of this exercise is to write a MATLAB function that achieves the following goals:

(a) Receives as its first input argument the full path to an MP3 file and checks if the file exists.

(b) If it exists, the FFmpeg tool is called to convert the MP3 file to a *temporary* WAVE file.

(c) It plots and plays the signal in the temporary file.

(d) It deletes the temporary file.

7. (D5) MATLAB provides the Graphical User Interface Development Environment (GUIDE) for the creation of Graphical User Interfaces (GUIs). In this exercise, you are asked to:

(a) Use GUIDE to create a simple user interface composed of: (a) a LOAD button, (b) a PLAY button, (c) a SAVE button, (d) two plotting areas, and (e) two sliders (one slider below each plotting area).

(b) When the LOAD button is pressed, a file selection dialog appears that lets the user select a WAVE file. To this end, use the uigetfile() function. If the selected WAVE file is in STEREO mode, then each channel is plotted separately in the respective plotting area. If not, the same signal is plotted on both areas. Furthermore, if the signal is in MONO mode, a STEREO signal is synthesized by using the same signal in both channels.

(c) The two sliders are used to amplify the respective channels. The amplification is achieved by simple multiplication of each channel with the factor defined by the slider. Each slider produces values in the range [0,5], with values <1 attenuating the signal and values >1 amplifying it.

(d) When the PLAY button is pressed, the signal is played. Furthermore, if the SAVE button is pressed then the signal is stored in a predefined WAVE file.

8. (D5) One way of loading video files in MATLAB is by creating objects of the `VideoReader` class. In this exercise you are asked to develop a MATLAB program that:

(a) Takes as input the path to a video file (AVI format), calls the FFmpeg tool to extract the audio signal from the video file and stores the signal into a WAVE file.

(b) Applies short-term processing (Section 2.7) on the audio signal and computes the average absolute value of the samples in each frame. The length of the moving window is equal to 100 ms with zero overlap between successive windows. The result of this stage is a sequence of values.

(c) Detects the maximum value of the previous sequence and assumes that the time instant, t_{max}, at which this value occurs coincides with the time equivalent of the center of the respective frame.

(d) Retrieves the video frame that corresponds to t_{max} and uses the `imshow()` function to display it.[3]

9. (D4) Center clipping is a useful technique according to which a signal sample survives if its absolute value exceeds a predefined threshold. Equation (2.3) defines the clipping operation when a single threshold, T_h, is used.

$$x_c(n) = \begin{cases} x(n) - Th, & \mid x(n) \mid \geq T_h, \\ 0, & \text{otherwise,} \end{cases} \quad (2.3)$$

where $x_c(n)$ and $x(n)$ are the clipped and original signal, respectively. In this exercise, you are asked to develop a function that receives as input a

---

[3] The `VideoReader` class may not be available for older versions of MATLAB. In which case, you can alternatively use the `mmread()` function, which is available in the Mathworks File Exchange website (http://www.mathworks.com/matlabcentral/fileexchange/8028-mmread).

monophonic audio signal and a clipping threshold and plots the clipped signal on a short-term processing basis. You may find it useful to reuse parts of the source code of function stpFile(). The length and step of the moving window are user-defined parameters.

10. (D5) The object of this exercise is to develop a MATLAB program that shuffles an audio signal as follows:

- Starts with a WAVE file and reads its contents in block-mode. The block length, B, is a user parameter.
- Stores each block as the row of a matrix, say X. Employs zero padding for the last block, if its length is shorter than B.
- After all blocks have been read, shuffles the rows of matrix X. To this end, function randperm can prove to be useful.
- Creates a new WAVE file by storing each row of X, one after the other, in this new file.
- Listens to the result.

Depending on the original file, shuffling can create an audio effect that can be used as a tapestry of sounds in the background of a recording. This is a well-known technique in music recording studios. Finally, create a stereo recording by placing the shuffled audio signal in the left channel and a signal of your choice in the right channel. Save the result and listen to it.

CHAPTER 3

Signal Transforms and Filtering Essentials

Contents

The goal of this chapter is to provide a gentle introduction to selected signal transforms, which generate representations that are useful in various audio-related tasks, including feature extraction, compression, and multiresolution analysis, to name but a few. We have placed special emphasis on the description of the Discrete Fourier Transform because a lot of material in later chapters of this book assumes that the reader is familiar with this particular transform. We also present the fundamentals of digital filters, so that, by the end of the chapter, the reader will be able to create and experiment with basic filter types and understand how they can affect the performance of various audio analysis stages.

3.1. THE DISCRETE FOURIER TRANSFORM

The *Discrete Fourier Transform (DFT)* is of paramount importance in all areas of digital signal processing. It is used to derive a frequency–domain (spectral) representation of the signal. As will be made clear in the next chapter on feature extraction, the majority of the important features used to analyze audio content are defined in the frequency domain. It is, therefore, important to gain a firm understanding of the DFT. To this end, we will focus on presenting the transform from an implementation perspective and our discussion will evolve around the interpretation of the output of the DFT and how it reflects the properties of the audio signal. Before we proceed, note that an efficient algorithm for the computation of the DFT coefficients is the Fast Fourier Transform (FFT), which exploits the computational redundancy in the equations that define the DFT and its inverse transform.

Given a discrete-time signal, $x(n)$, $n = 0, \ldots, N - 1$, N samples long, its DFT is defined as:

$$X(k) = \sum_{n=0}^{N-1} x(n) \exp\left(-j\frac{2\pi}{N}kn\right), \quad k = 0, \ldots, N - 1, \tag{3.1}$$

where $j \equiv \sqrt{(-1)}$. It can be observed that the output of the transform is a sequence of N coefficients, the $X(k)s$, which, in general, are complex numbers. The DFT coefficients constitute the frequency-domain representation of the signal, which is further explained later in this chapter.

The inverse DFT (IDFT) takes as input the DFT coefficients and returns the original signal:

$$x(n) = \frac{1}{N} \sum_{k=0}^{N-1} X(k) \exp\left(j\frac{2\pi}{N}kn\right), \quad n = 0, \ldots, N - 1. \tag{3.2}$$

Equation (3.2) provides an exact reconstruction of the original signal. As a result, the time-domain signal, $x(n)$, $n = 0, \ldots, N - 1$ and the complex signal, $X(k)$, $k = 0, \ldots, N - 1$ can be treated as equivalent representations. If we rewrite Eq. (3.2) in the form

$$x(n) = \frac{1}{N} \sum_{k=0}^{N-1} X(k)\gamma_k(n), \quad n = 0, \ldots, N - 1, \tag{3.3}$$

where $\gamma_k(n) = \exp\left(j\frac{2\pi}{N}kn\right)$, $n = 0, \ldots, N - 1$, then it can be seen that the original signal, $x(n)$, can be written as a weighted average of a family of fundamental signals (basis functions), where each signal, $\gamma_k(n)$, is a complex exponential and its weight is equal to the kth DFT coefficient.

A useful interpretation of the DFT coefficients, in which we are particularly interested for practical reasons, is that, if F_s is the sampling frequency that was used to obtain $x(n)$, then the kth exponential corresponds to the analog frequency $f_k = k\frac{F_s}{N}$, $k = 0, \ldots, N - 1$ (the discrete-time equivalent of which is $\omega_k = k\frac{2\pi}{N}$). For a given sampling frequency, larger values of N (i.e. longer signals) result in a more dense sampling of the frequency axis, because $\frac{F_s}{N}$ becomes smaller. Therefore, a larger value of N is expected to produce a finer representation of the signal in the frequency domain, i.e. a better *frequency resolution*. However, there is a subtle issue here, which we will demonstrate later on by means of certain examples: the whole discussion on increasing frequency resolution is valid, as long as the signal remains *stationary*, i.e. as long as its properties do not change over time. For example, a sum of sinusoids of very long duration is a stationary signal, which can be easily studied with high-frequency resolution. On the other hand, if a signal changes from a sum of two sinusoids to a sum of three sinusoids, it is no longer stationary. It can only be considered stationary on a local basis. In such situations, N cannot increase arbitrarily. As a consequence, we cannot always achieve the desirable frequency resolution, because N is constrained by the stationarity changes in the signal. In other words, we need to resort to *a trade-off between frequency and time resolution*.

Another discussion in which we are particularly interested involves the magnitude of the kth DFT coefficient, $| X(k) |$, which can be treated as a measure of the *intensity* with which the respective frequency participates in the signal $x(n)$. This leads us to the aforementioned spectral interpretation of the DFT. Note that the phase of the DFT coefficients can also play a useful role in various applications and its role should not be underestimated. However, the majority of feature extraction methods are heavily based on the magnitude of the DFT coefficients.

MATLAB has adopted the FFTW open-source library [6], http://www.fftw.org, for the implementation of the DFT. The respective built-in function is fft(). To compute the DFT of a one-dimensional signal, which is stored in the MATLAB vector x, type

```
X = fft(x);
```

By the definition of the DFT, the length of vector X is equal to the length of the input signal. An engineering trick that is often used in association with the computation of the DFT is the *zero-padding* technique. The goal of zero padding is to achieve increased frequency resolution to the expense of adding a number of zeros in the end of the input signal. To enable zero padding, call the fft() function as follows:

```
X = fft(x, N);
```

where N is the desired number of DFT coefficients, after zero padding has taken place. If the input signal is longer than N samples then X is truncated.

The magnitude of the DFT coefficients can be easily computed by means of the abs() MATLAB function. As an example, type:

```
x = [−1 2 0 −1 1 2 3 2 1];
X = fft(x);
magX = abs(X);
```

The vector of the resulting DFT coefficients is

```
9.0, −2.17+5.13i, −1.06−4.41i, −3.0−1.73i, −2.77+0.85i,
−2.77−0.85i, −3.0+1.73i, −1.06+4.41i, −2.17−5.13i
```

Consequently, the output of the abs() function is

`9.0, 5.57, 4.53, 3.46, 2.89, 2.8942, 3.46, 4.53, 5.57`

If we take a closer look at the DFT coefficients and their magnitude, we can verify a well-known property that is derived from the theory of the DFT: for a real-valued signal, the DFT coefficients appear in conjugate pairs, i.e. $X(k) = \overline{X}(N - k), k = 1, \ldots, N - 1$. This implies that the magnitude of the spectrum is symmetric. In the above example, $N = 9$ and: $\mid X(1) \mid = \mid X(N - 1) \mid = 5.57, \mid X(2) \mid = \mid X(N - 2) \mid = 4.53$, etc. The first DFT coefficient, $X(0)$, is known as the DC component of the signal and it is equal to the sum of signal samples. In this example, $X(0) = \sum_{n=0}^{N-1} x(n) = 9$.

Due to the symmetric property of the magnitude of the DFT coefficients, we only need the DFT coefficients with indices $k = 0, \ldots, \lceil \frac{N-1}{2} \rceil$, where $\lceil \cdot \rceil$ is the ceiling operator. In the previous example, we only need the first five coefficients ($k = 0, \ldots, 4$). If we continue this line of thinking, we understand that although the frequencies of the DFT coefficients cover the range $[0 - (N - 1)\frac{F_s}{N}]$ Hz, in practice, we only need the frequencies up to $\frac{F_s}{2}$, which is also in agreement with the Nyquist theorem.

If the concept of negative frequency is introduced, it can be equivalently stated that the DFT frequencies cover the range $[-\frac{F_s}{2}, +\frac{F_s}{2}]$, which can be useful for visualization purposes. In the following example, (getDFT()), this is achieved by a call to the fftshift() function, which is fed with the DFT coefficients of the signal. The getDFT() function returns the magnitude of the DFT coefficients, along with the corresponding frequencies (in Hz). Its standard input arguments are the signal and the sampling rate. If a third argument is provided, the function additionally returns the DFT coefficients arranged in the range $[-\frac{F_s}{2}, \frac{F_s}{2}]$, otherwise, only the 'positive' part of the spectrum is returned.

```
function [FFT, Freq] = getDFT(signal, Fs, PLOT)


%
% function [FFT, Freq] = getDFT(signal, Fs, PLOT)
%
% This function returns the DFT of a discrete signal and the
% respective frequency range.
%
% ARGUMENTS:
% - signal: vector containing the samples of the signal
% - Fs:     the sampling frequency
% - PLOT:   use this argument if the FFT (and the respective
%           frequency values) need to be returned in the
%           [-fs/2..fs/2] range. Otherwise, only half of
%           the spectrum is returned.
%
```

```
% RETURNS:
% — FFT:     the magnitude of the DFT coefficients
% — Freq:    the corresponding frequencies (in Hz)
%


N = length(signal);  % length of signal
% compute the magnitude of the spectrum
% (and normalize by the number of samples):
FFT = abs(fft(signal)) / N;

if nargin==2 % return the first half of the spectrum:
    FFT = FFT(1:ceil(N/2));
    Freq = (Fs/2) * (1:ceil(N/2)) / ceil(N/2);  % define the frequency axis
else
    if (nargin==3)
        % ... or return the whole spectrum
        %     (in the range —fs/2 to fs/2)
        FFT = fftshift(FFT);
        if mod(N,2)==0                           % define the frequency axis:
            Freq = —N/2:N/2—1;                   % if N is even
        else
            Freq = —(N—1)/2:(N—1)/2;             % if N is odd
        end
        Freq = (Fs/2) * Freq ./ ceil(N/2);
    end
end
```

The fftExample() function demonstrates how to call getDFT() by generating a sum of sinusoidal signals with frequencies provided by the user. In the body of the fftExample() m-file, function getDFT() is called twice to compute the spectrum of the generated signal in the two frequency ranges that were described earlier. The time duration of the resulting signal is provided as an input argument to the function.

```
function fftExample(Fs, f, duration)


% function fftExample(Fs, f, duration)
%
% Demonstrates the use of the getDFT() function.
% Generates a sum of sinusoidal signals and computes—plots its DFT amplitude.
%
% ARGUMENTS:
% — Fs: sampling frequency
% — f:  frequencies of the tones (the signal is generated
%       as a sum of sinusoidal signals)
% — duration: the duration of the signal (in seconds)


t = 0:1/Fs:duration;               % time vector
x = cos(2*f(1)*pi*t);              % create the signal
for (i=2:length(f)) x = x + cos(2*f(i)*pi*t);  end
x = x / length(f);                 % signal normalization


% compute the magnitude of the spectrum:
```

```
[X, FreqX] = getDFT(x, Fs);          % freq range: 0–>fs/2
[X2, FreqX2] = getDFT(x, Fs, 1);     % freq range: –fs/2–>fs/2
% plot the results:
figure; subplot(2,1,1); plot(FreqX, X, ' k'); title('Magnitude of DFT');
xlabel('Hz'); title('Positive part of spectrum');
subplot(2,1,2); plot(FreqX2, X2, ' k'); title('Magnitude of DFT');
xlabel('Hz'); title('Spectrum in range – f_s / 2 –> f_s / 2');
```

To call the `fftExample()` function, type:

```
fftExample(8000, [200 500 1200], 0.5)
```

The code generates a signal consisting of three tones at 200, 500, and 1200 Hz. The sampling rate is 8 kHz and the signal duration is 0.5 s. The output of the function is plotted in Figure 3.1. The peaks of the magnitude of the spectrum correspond to the three frequencies of the signal.

The next example (`fftSNR()`) demonstrates the impact of noise on the spectrum of a signal. The function starts by generating a sum of sinusoids as it was done in the previous example. At a second step, Gaussian noise is added to the signal. The signal-to-noise ratio (SNR) is approximately 15 dB and is a user-defined parameter. In the end, the magnitude of the spectrum of both the clean and noisy signals are computed and plotted in decibels.

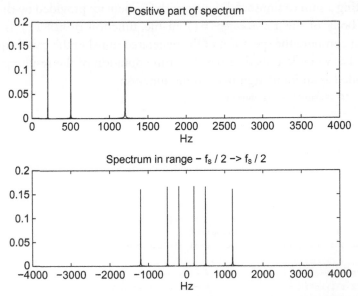

Figure 3.1 Plots of the magnitude of the spectrum of a signal consisting of three frequencies at 200, 500, and 1200 Hz. The sampling frequency is 8 kHz. The frequency ranges on the horizontal axes are [−4 to +4] kHz and [0–4] kHz, respectively.

```
function fftSNR(Fs, f, duration, SNR)


%
% function fftSNR(Fs, f, duration, SNR)
%
% ARGUMENTS:
% — Fs: sampling frequency
% — f:  frequencies of the tones (the signal is generated as sum of
%       sinusoidal signals)
% — duration: the duration of the signal (in seconds)
% — SNR: the signal to noise ratio (in dBs)
%


t = 0:1/Fs:duration;                  % time vector
% signal definition:
x = cos(2*f(1)*pi*t);                 % a. clean signal:
for (i=2:length(f)) x = x + cos(2*f(i)*pi*t);  end
x = x / length(f);                    % signal normalization
y = awgn(x, SNR, 'measured');         % b. noisy signal:


% compute the magnitude of the spectrum of x and y:
[X, FreqX] = getDFT(x, Fs); [Y, FreqY] = getDFT(y, Fs);


% plot the results:
figure; subplot(2,1,1); plot(FreqX, log10(X), 'k');
axis([1 max(FreqX) —5 0]);
title('Log—magnitude of the Spectrum of the original (clean) signal');
xlabel('Frequency (Hz)');
subplot(2,1,2); plot(FreqY, log10(Y), 'k');
axis([1 max(FreqX) —5 0]);
title('Log—magnitude of the Spectrum of the noisy signal');
xlabel('Frequency (Hz)');
```

To call fftSNR() type:

```
fftSNR(8000, [200 500 1200], 0.5, 15)
```

The output of this example is plotted in Figure 3.2.

3.2. THE SHORT-TIME FOURIER TRANSFORM

The goal of the Short–Time Fourier Transform (STFT) is to break the signal into possibly overlapping frames using a moving window technique and compute the DFT at each frame. Therefore, the STFT falls in the category of short–term processing techniques (see also Section 2.7). As has already been explained, the length of the moving window plays an important role because it defines the frequency resolution of the spectrum, given the sampling frequency. Longer windows lead to better frequency resolution at the expense of decreasing the quality of time resolution. On the other hand, shorter windows provide a more detailed representation in the time domain, but, in general, lead to poor frequency resolution. In audio analysis applications, the

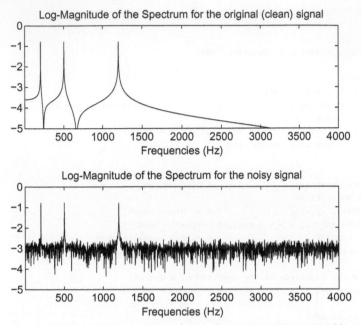

Figure 3.2 A synthetic signal consisting of three frequencies is corrupted by additive noise. The spectra of the original and the noisy signals are plotted (the SNR is 15 dB).

short-term window length usually ranges from 10 to 50 ms. The MPEG7 audio standard recommends that it is a multiple of 10 ms [7].

If the DFT coefficients of each frame are placed into a separate column of a matrix, the STFT can be represented as a matrix of coefficients, where the column index represents time and the row index is associated with the frequency of the respective DFT coefficient. If the magnitude of each coefficient is computed, the resulting matrix can be treated as an image and, as a result, it can be visualized. This image is known as the *spectrogram* of the signal and presents the evolution of the signal in the time-frequency domain. To generate the spectrogram, we can use the magnitude or the squared magnitude of the STFT coefficients on a linear or logarithmic scale (dB). In MATLAB, the spectrogram of a signal is implemented in the `spectrogram()` function, which can plot the spectrogram and return the matrix of STFT coefficients, along with the respective time and frequency axes. In this book, we will mainly use the spectrogram as a visualization tool. The STFT coefficients will be extracted, when required, by means of a more general function that we have developed for short-term processing purposes. In Figure 3.3, we present an example of a spectrogram for a speech signal.

It is also important to note that before the computation of the DFT coefficients takes place, each frame is usually multiplied on a sample basis

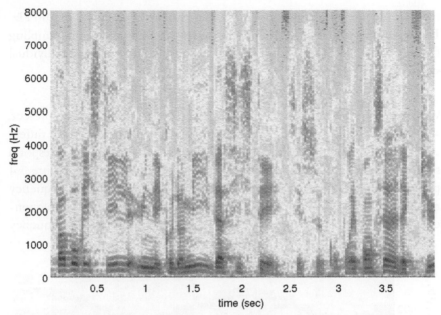

Figure 3.3 The spectrogram of a speech signal. The frames are non-overlapping, 20 ms long.

with a window function, which aims at attenuating the endpoints of the frame while preserving the signal around the center of the frame. Popular windows have been implemented by MATLAB in the hamming(),hann(), and blackman() m-files. Each function creates a symmetric window based on a different formula.

We are now revisiting the issue of frequency resolution with an example that revolves around the STFT of two frequency-modulated signals. Before we proceed with the example, remember that, on a frame basis, the distance (in Hz) between two successive DFT coefficients is equal to $\frac{F_s}{N}$ Hz, where F_s is the sampling frequency and N is the number of samples of the frame. The frequency resolution determines when two close frequencies will be distinguishable in the spectrum of the signal. For a fixed sampling rate, F_s, in order to improve the DFT resolution, we need to increase the length of the short-term frame. However, as it has already been explained, the price to pay is decreased time resolution. Our example creates the following two synthetic signals:

$$x_1(n) = cos(2\pi \cdot 500n + 200cos(2\pi \cdot n)),$$
$$x_2(n) = cos(2\pi \cdot 590n + 200cos(2\pi \cdot n)).$$

Each signal is *frequency modulated*. Consider for example the first equation: the term $200cos(2\pi \cdot n)$ creates a 1 Hz signal modulation, which has the effect that the frequency of the tone at 500 Hz exhibits a variation of 200 Hz. The resulting synthetic signal is 2 s long. In this experiment, the sampling frequency is 2 kHz. At the final stage, we create the sum of the two individual signals:

$$x(n) = x_1(n) + x_2(n).$$

We therefore expect that, at each frame, two frequencies should be present, separated by 90 Hz. We now generate the spectrogram of x for three different short-term frame lengths, namely 100, 50, and 10 ms, which correspond to the frequency resolutions of 10, 20, and 100 Hz, respectively. Figure 3.4 presents the generated spectrograms. In the third spectrogram, the frequency resolution (100 Hz) is incapable of distinguishing between the two frequencies, which appear as a single wide band. On the contrary, the frequency

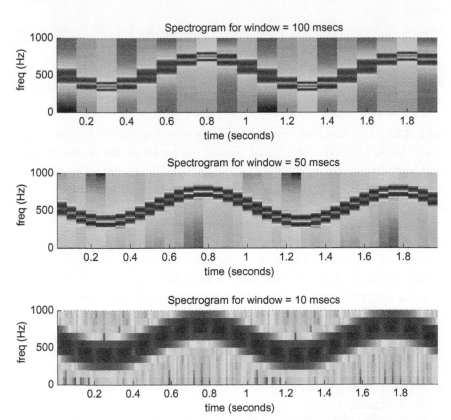

Figure 3.4 Spectrograms of a synthetic, frequency-modulated signal for three short-term frame lengths. The best result is achieved by the second spectrogram, because it provides a good trade-off between frequency and time resolution. On the contrary, the first short-term frame length leads to low time resolution, while the third frame length decreases frequency resolution.

resolution of 10 Hz (first spectrogram) separates the two frequencies (which differ by ≈90 Hz), *but* it fails to follow accurately the time evolution of the signal. It turns out that the best choice is the 50 ms frame length, which provides the best trade-off between frequency and time resolution.

3.3. ALIASING IN MORE DETAIL

The aliasing effect was introduced in Section 2.1. We are now revisiting the aliasing phenomenon with an example involving synthetic signals and through the adoption of frequency representations. Remember that aliasing occurs when the sampling rate is insufficient with respect to the frequency range of the signal, and that, according to the Nyquist theorem, the sampling frequency must be at least twice the highest frequency of the signal to be sampled. From a theoretical point of view, the process of sampling an analog signal creates a discrete-time signal whose frequency representation is a superposition of an infinite number of copies of the Fourier transform of the original signal, where each copy is centered at an integer multiple of the sampling frequency [8–10]. This is presented in Figure 3.5. The second

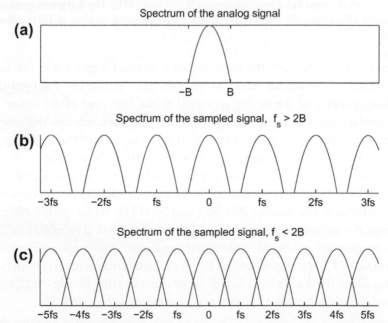

Figure 3.5 Spectrum representations of (a) an analog signal, (b) a sampled version when the sampling frequency exceeds the Nyquist rate, and (c) a sampled version with insufficient sampling frequency. In the last case, the shifted versions of the analog spectrum are overlapping, hence the aliasing effect.

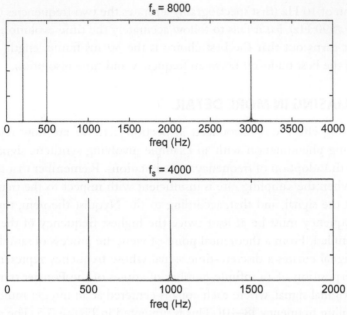

Figure 3.6 Spectral representations of the same three-tone (200, 500 and 3000 Hz) signal for two different sampling frequencies (8 kHz and 4 kHz). The 4 kHz rate generates an aliasing effect because the 3KHz frequency vanishes and its alias at 1KHz appears instead.

part of the figure displays the spectrum of a sampled signal when the sampling frequency is higher than the Nyquist rate. It can be observed that the shifted copies of the analog spectrum of the first part of the figure are non-overlapping. On the contrary, the third part demonstrates the case of insufficient sampling rate. It can be seen that the replicas of the analog spectrum are now overlapping, giving rise to the aliasing phenomenon.

Figure 3.6 presents the aliasing phenomenon for a simple signal. The signal $x(t) = cos(2\pi \cdot 200t) + cos(2\pi \cdot 500t) + cos(2\pi \cdot 3000t)$ consists of a sum of three tones at frequencies 200, 500, and 3000 Hz. Based on the Nyquist theorem, the minimum acceptable sampling rate is $3000 \cdot 2 = 6000$ Hz. The figure presents the spectra of the sampled signal when the sampling frequency is 8 kHz and 4 kHz, respectively. The second sampling frequency causes an aliasing effect: the 3 kHz tone fails to be preserved in the discrete–time signal representation and its alias appears as a new frequency at 1000 Hz.

3.4. THE DISCRETE COSINE TRANSFORM

Another transform that has been widely used in the field of signal processing is the Discrete Cosine Transform (DCT) [11,5]. It is a purely real transform, because, unlike the DFT, which uses complex exponentials as its basis functions, the DCT expresses a finite duration signal as a weighted sum of cosines (real functions). The DCT has been very popular in the development of coding applications, especially in the case of two-dimensional signals (images).

Let $x(n)$, $n = 0, \ldots, N - 1$ be a discrete-time signal of finite duration. Its DCT is defined as

$$y(k) = a(k) \sum_{n=0}^{N-1} x(n) cos\left(\frac{\pi(2n+1)k}{2N}\right), \quad k = 0, 1, \ldots, N - 1. \quad (3.4)$$

We observe that Eq. (3.4) generates N DCT coefficients, $y(k)$, $k = 0, 1, \ldots, N - 1$, which will serve as the weights of the basis functions in the following equation that defines the inverse DCT:

$$x(n) = \sum_{k=0}^{N-1} a(k) y(k) cos\left(\frac{\pi(2n+1)k}{2N}\right), \quad n = 0, 1, \ldots, N - 1, \quad (3.5)$$

where

$$a(k) = \begin{cases} \sqrt{\frac{1}{N}}, & k = 0, \\ \sqrt{\frac{2}{N}}, & k \neq 0. \end{cases} \quad (3.6)$$

An interesting property of the DCT, which leads to computationally efficient implementations, is that it can be expressed as a product of matrices, i.e. $y = C^T x$, where C has real elements and $C^{-1} = C$. In the MATLAB programming environment, the DCT has been implemented in the dct() function of the Signal Processing Toolbox.

Another interesting feature of the DCT is that it provides very good information—packing options. This is why the two-dimensional DCT has been embedded in the JPEG image compression standard. The one-dimensional DCT has also been adopted by audio coding standards. For example, the MPEG1-Layer III (MP3) standard uses a modified DCT. Further to its adoption by compression techniques, the DCT has been also employed by various feature extraction algorithms. An outstanding example is the adoption of the inverse DCT as the last processing stage in the extraction of the Mel-Frequency Cepstrum Coefficients (MFCCs) from the audio signal [8]. More details on this matter can be found in Section 4.4.5, in the following chapter.

In this section, we will present an example that demonstrates the basic ideas behind the DCT-based compression and decompression of an audio signal. Our simple compression-decompression scheme goes through the following steps:

- The signal is split into non-overlapping windows. The length of the moving window is set equal to 100 ms.
- For each window, the DCT is computed.
- The DCT coefficients are sorted in descending order based on their amplitude. The DCT coefficients which correspond to the CR% of highest amplitude values are preserved along with their indices. The remaining DCT coefficients are set to zero.
- The inverse DCT is applied on the modified sequence of DCT coefficients. The result is a real-valued, time-domain signal, which has preserved the dominant characteristics of the original signal and has discarded unnecessary detail.

The dctCompress() function implements the compression steps of the above scheme. The DCT coefficients that survive at each frame along with the respective indices are stored as rows of data in two separate matrices. To decompress the signal, the inverse procedure is executed: the DCT coefficients at each row of the respective matrix are used to compute the inverse DCT. The decompression stage has been implemented in the dct Decompress() function. The following piece of code demonstrates how to use functions dctCompress() and dctDecompress(), given an input signal. Part of the example file diarizationExample.wav has been used here, which is also included in the library that comes with this book. The only input argument of this demo function is the required compression ratio.

```
function demoDCTcompression(cR)
% Demonstrates how to use the DCT audio compression and decompression
% functions

% load audio signal from a sample wav file
[x,fs] = wavread([['..' filesep 'data' ...
    filesep 'diarizationExample.wav'], [280000 300000]);
% generate compressed DCT coefficients and indices
[DCTcoeffs, INDcoeffs] = dctCompress(x, 0.10, fs, cR);
% decompress using iDCT:
x2 = dctDecompress(DCTcoeffs, INDcoeffs, 0.10, fs);
% listen to the results:
fprintf('Listening to the original sound ...');sound(x, fs);
fprintf('\nListening to the decompressed sound (%.2f ratio)...\n', cR);
sound(x2, fs);
```

3.5. THE DISCRETE-TIME WAVELET TRANSFORM

The goal of the Discrete-Time Wavelet Transform (DTWT) is to provide a multiresolution analysis framework for the study of signals. The term multiresolution refers to the fact that the signal is split into a hierarchy of signal versions of increasing detail. The original signal can be reconstructed from its versions by means of a weighting scheme, which is defined by the inverse DTWT. The theory of the DTWT is quite complicated and can be approached from different perspectives. For example, it is common to present the DTWT from the point of view of tree-structured filter banks, which in turn requires that the reader is familiar with filter theory. In this section we will only provide the analysis and synthesis equations of the DTWT and its inverse, following the presentation that we also adopted for the DFT and the DCT. We will comment on the respective equations and then we will explain the steps of multiresolution analysis using MATLAB code.

To begin with, the DTWT is defined as

$$y_i(k) = \sum_n x(n)\phi_{ik}(n), \quad i = 0, 1, \ldots, J - 1, \tag{3.7}$$

where index i represents the analysis (resolution) level, k is the index of the DTWT coefficients at the ith level, and the $\phi_{ik}(n)$s are appropriately defined functions. The inverse DTWT performs a perfect reconstruction of the original signal and it is given by the equation

$$x(n) = \sum_i \sum_k y_i(k)\psi_{ik}(n), \tag{3.8}$$

where the $\psi_{ik}(n)$s are the basis functions at the ith level. An important issue is that the functions at each analysis level stem from an appropriately defined mother function:

$$\phi_{ik}(n) = \phi_{i0}(n - 2^r k), \tag{3.9}$$

where $r = i + 1$ if $i \neq J - 1$, $r = J - 1$ if $i = J - 1$ and $\phi_{i0}(n)$ is the mother sequence of the ith level. It can be seen that all the functions of an analysis level are shifted versions of the mother function. Similarly, for the inverse transform,

$$\psi_{ik}(n) = \psi_{i0}(n - 2^r k), \tag{3.10}$$

where $r = i + 1$ if $i \neq J - 1$, $r = J - 1$ if $i = J - 1$ and $\psi_{i0}(n)$ is the mother sequence of the ith level. You have probably guessed that if the reconstruction of the original signal takes into account less than J levels, then an approximate (coarse) version of the original signal will be made available.

In order to get a better understanding of the DTWT and its inverse, let us follow a hierarchical decomposition of a signal using a series of MATLAB commands. MATLAB provides the functionality that we need in the Wavelet Toolbox. Specifically, we will use the dwt() function for splitting the signal into two components and the `idwt()` function for the reconstruction stage. First of all, we need to load a signal. To this end, we select a segment, 1024 samples long, drawn from the previously used `diarizationExample.wav` file.

```
[x,fs] = wavread(['..' filesep 'data' ...
         filesep 'diarizationExample.wav'], [280000 281023]);
```

Then we decompose the signal into two components:

```
[ca,cd] = dwt(x,'db1');
```

The second input argument specifies the mother function of the analysis stage. The first output argument, *ca*, is the vector of DTWT coefficients of the coarse analysis level and the second output argument, *cd*, contains the coefficients, which capture the detail of the signal. Note that the length of each output signal is half the length of the input signal *x*. We can now plot the two signals:

```
clf;figure(1)
subplot(211),plot(ca); title('Approximation coefficients')
subplot(212),plot(cd); title('Detail coefficients')
```

The next step is to further decompose the coarse approximation, *ca*, into two signals, *caa* and *cad*:

```
[caa,cad] = dwt(ca,'db1');
```

This time, vectors *caa* and *cad* contain 256 analysis coefficients. We can now plot *caa*, *cad*, and *cd* in a single figure:

```
clf;figure(1)
subplot(311),plot(caa);
subplot(312),plot(cad);
subplot(313),plot(cd);
```

In other words, we have performed a two-stage analysis of the original signal. We will now proceed with the inverse operation, i.e. we will reconstruct the signal in two stages. For the first reconstruction stage, type:

```
x1=idwt(caa,cad,'db1');
```

The reconstructed signal, *x*1, is 512 samples long. To proceed with the next reconstruction stage, type:

```
x_hat=idwt(x1,cd,'db1');
```

Note that if we use zeros instead of the *cd* signal, this will mean that we are not interested in the detail that the *cd* signal carries. If this is desirable, we can type:

```
x_hat=idwt(x1,zeros(length(cd),1),'db1');
```

If you plot *x_hat* and the original signal in the same figure, you will readily observe that their differences are really minor. In practice, this means that we were right to ignore the detail in the *cd* signal. The respective SNR (signal-to-noise ratio) will reveal the same observation (see also Exercise 7, which is on the computation of SNR).

3.6. DIGITAL FILTERING ESSENTIALS

The goal of this section is to provide a gentle introduction to the design of digital filters, focusing on reproducible MATLAB examples. Digital filters are important building blocks in numerous contemporary processing systems that perform diverse operations, ranging from quality enhancement to multiresolution analysis. The interested reader may consult several textbooks in the field, which provide a detailed treatment of the subject from both a theoretical and implementation perspective, e.g. [8,1,12,13].

To begin with, consider the following equation, which describes the signal $y(n)$ at the output of a system as a function of the signal $x(n)$ at its input:

$$y(n) = x(n) + a \cdot x(n-1), \tag{3.11}$$

where a is a constant and $\mid a \mid < 1$. Equation (3.11) describes a *filter*, i.e. a system whose output at the nth discrete-time instant is computed as a function of the current input sample, $x(n)$ and a weighted version of the previous input sample, $x(n-1)$. Now, let us assume that the following signal is given as input to the filter:

$$\delta(n) = \begin{cases} 1, & n = 0, \\ 0, & \text{otherwise.} \end{cases} \tag{3.12}$$

This signal is also known as the *unit impulse sequence*. It is straightforward to compute the output of the system for this simple signal:

$$h(n) = \begin{cases} 1, & n = 0, \\ a, & n = 1, \\ 0, & \text{otherwise,} \end{cases} \tag{3.13}$$

where $h(n) \equiv y(n)$. The response of the system to the unit impulse sequence is known as the *impulse response* of the system. The impulse response is a signal that offers an alternative representation of the system. In this case, the impulse response becomes zero after the second sample ($n = 1$). When this type of behavior is observed, we are dealing with a *Finite Impulse Response (FIR) filter* (system). In the more general case, a FIR system is described by the equation:

$$y(n) = \sum_{k=0}^{N} b(k)x(n-k) = \sum_{k=0}^{N} h(k)x(n-k). \tag{3.14}$$

In our example,

$$y(n) = h(0) \cdot x(n) + h(1) \cdot x(n-1). \tag{3.15}$$

A quick inspection of Eq. (3.14) reveals that the output of a FIR system is always a function of the current and past samples of the input signal. Number N is known as the order of the filter.

The DFT of the impulse response is called the *frequency response of the filter*. Given that the impulse response of the FIR filters becomes zero after a finite number of samples, it follows that, if we want to study the frequency response of a FIR system with sufficient frequency detail, we need to resort to the technique of zero padding that was described in the context of the DFT transform.

Assuming that our filter operates at a sampling frequency of 1000 Hz, we can type the following commands to generate and plot the magnitude of its frequency response:

```
a=-0.95;
h=[1 a];
H=fft(h,1024); % padded with 1022 zeros
Fs=10000; % sampling frequency
N=length(H);
k=0:N-1; fk=k*Fs/N;
plot(fk,20*log10(abs(H))); % plot the magnitude of the frequency response in dB
```

In the above lines of code the a parameter was set equal to -0.95. As a result, on a dB scale, the frequency response is negative for the low–frequency range [0–1700] Hz (approximately) and positive for frequencies higher than 1700 Hz, as shown in Figure 3.7. In other words, the filter attenuates the low–frequency range and amplifies the frequencies in the range [1700–5000] Hz. This type of frequency response is particularly useful as a preprocessing stage when the human voice is recorded over a microphone, because most micro-phones tend to emphasize low frequencies. Our filter has the inverse effect: it emphasizes the high frequencies (if $a < 0$), hence the name *pre-emphasis filter*.

FIR systems fall in the broader category of *Linear Time Invariant (LTI)* systems. The linear property defines that a linear combination of inputs will produce the same linear combination of outputs. For example, if our simple filter is fed with the signal $z(n) = c \cdot x_1(n) + d \cdot x_2(n)$, the output will be:

$$
\begin{aligned}
y(n) &= z(n) + a \cdot z(n-1), \\
&= c \cdot x_1(n) + d \cdot x_2(n) + a(c \cdot x_1(n-1) + d \cdot x_2(n-1)), \\
&= c \cdot y_1(n) + d \cdot y_2(n),
\end{aligned}
$$

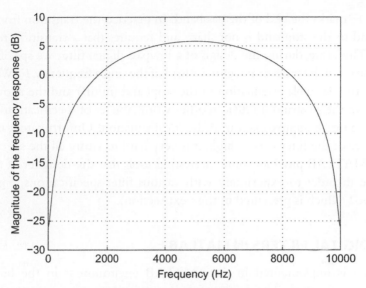

Figure 3.7 Frequency response of a pre-emphasis filter for $a = -0.95$.

where $y_1(n)$ and $y_2(n)$ are the signals at the output of the filter when the input signals are $x_1(n)$ and $x_2(n)$, respectively. The property of time invariance defines that a shifted version of the input will produce a shifted version of the output by the same amount of time shift.

A different category of filters can be designed if the right side of the input–output equation includes past values of the output:

$$y(n) = \sum_{k=1}^{N} a_k y(n-k) + \sum_{m=0}^{M} b_m x(n-m). \tag{3.16}$$

If at least one a_k is nonzero, Eq. (3.16) describes the category of *Infinite Impulse Response (IIR)* filters (otherwise it reverts to a description of a FIR filter). Due to the infinite length of their impulse response, we need the recurrence relation of Eq. (3.16) to compute the output of IIR filters, i.e. it no longer makes sense to use a summation like the one in Eq. (3.14).

Another important categorization of digital filters is based on the shape of the frequency response. Specifically, the filters can be categorized in the *lowpass, bandpass, highpass,* and *stopband* categories. Basically, the names indicate the frequency range (passband) where the magnitude of the frequency response of the filter is expected to be equal to unity (on a linear scale) or 0 dB (on a logarithmic scale). For example, the term lowpass means that the filter is expected to leave unaltered all frequencies in the range $[0–F_c]$ Hz, where F_c marks the end of the passband. In practice, the transition from the passband to the stopband is not sharp and requires that a transition region exists. Therefore, during the design of a lowpass digital filter we specify the boundaries of the transition region instead of the single frequency, F_c. We also specify the desirable attenuation in the stopband (in dB) and the allowable change in the passband (in dB). In other words, the set of specifications aims at providing an approximation of the ideal situation. Unfortunately, unless the order of the filter is very high, it is very hard to satisfy all the specifications. MATLAB provides some interesting tools that the designer of filters can use in order to experiment with various filter specifications (e.g. the `fdatool` which is presented in the next section).

3.7. DIGITAL FILTERS IN MATLAB

Filtering is implemented in the MATLAB environment in the built-in `filter()` function. The syntax of the function is:

```
y = filter(b, a, x)
```

The input arguments are:

1. x: a vector with the samples of the input signal.
2. b: a vector with the weights of the input samples $x(n-m)$, $m = 0, \ldots, M$, as in Eq. (3.16).
3. a: a vector that starts with '1' followed by the weights of $y(n - k)$, $k = 1, \ldots, N$, as in Eq. (3.16). If we are dealing with a FIR filter, we simply set $a = [1]$.

For example, to filter a signal that is stored in vector x through the pre-emphasis filter $y(n) = x(n) + a \cdot x(n - 1)$, type:

```
y = filter([1 a],[1],x);
```

Another useful function, with input arguments that follow the rationale of the `filter()` function is the `freqz()` function, which generates the frequency response of a filter, given the coefficients of Eq. (3.16). For example, the following commands generate and plot the frequency response of our pre-emphasis filter:

```
a=-0.95;
h=[1 a];
Fs=10000;
[H,F] = freqz(h,1,1024,Fs);
plot(F,20*log10(abs(H)));
```

The `fdatool` of the Signal Processing Toolbox of MATLAB provides a graphical user interface (GUI) to facilitate the design and analysis of digital filters. The user is allowed to describe the frequency response of the filter via a set of specifications. After the filter has been designed, the `fdatool` can

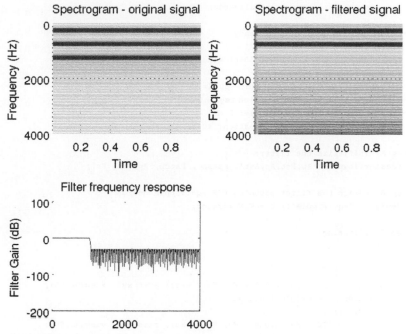

Figure 3.8 An example of the application of a lowpass filter on a synthetic signal consisting of three tones. The spectrogram of the output of the filter reveals that the highest tone has been filtered out.

be used to export the resulting filter coefficients (i.e. the a and b vectors) to the MATLAB workspace, to a MAT file, or to MATLAB code that can reproduce the filter design.

An alternative way to design a digital filter in MATLAB is by using the fdesign object of the Signal Processing Toolbox. Similar to the fdatool, the user again provides the filter specifications, i.e. type of filter, the critical frequencies and respective attenuations, etc.

The following example begins with the generation of a synthetic signal that consists of three frequencies. Then, it creates a *lowpass* filter and applies it on the generated signal. The spectrograms of the initial and the filtered signal, along with the frequency response of the designed filter are shown in Figure 3.8.

```
duration = 1; Fs = 8000; f = [200 600 1500];
t = 0:1/Fs:duration; % time vector
% signal definition:
x = cos(2*f(1)*pi*t);
for (i=2:length(f))
    x = x + cos(2*f(i)*pi*t);
end
```

```
x = x / length(f); % signal normalization

Fs = 8000;       % sampling frequency
t = 0: 1/Fs: 1; % time range
x = cos(2*pi*250*t) + cos(2*pi*750*t) + cos(2*pi*1250*t); % signal

Fpass = 1000;     % end of passband: 150 Hz
Fstop = 1050;     % end of transition region: 200 Hz
Ap = 0.5;         % band-pass ripple: 2 db
As = 30;          % stopband attenuation: 40 db

% Step A: initialize the filter:
d = fdesign.lowpass('Fp,Fst,Ap,Ast', Fpass , Fstop, Ap, As, Fs);

% Step B: design the filter assuming FIR equiripple design
F = design(d,'equiripple'); b = F.Numerator;

% Step C: filtering:
y = filter(b, 1, x);

subplot(2, 2, 1);
spectrogram(x, 0.020 * Fs, 0.010 * Fs, Fs/2, Fs); axis xy;  view(90,90);
title('Spectrogram — original signal');
subplot(2, 2, 2);
spectrogram(y, 0.020 * Fs, 0.010 * Fs, Fs/2, Fs); axis xy;  view(90,90);
title('Spectrogram — filtered signal');
% get frequency response of the designed filter:
[H, F] = freqz(b, 1, Fs, Fs);
subplot(2, 2, 3); plot(F, 20*log10(H));
xlabel('Frequency (Hz)'); ylabel('Filter Gain (dB)');
title('Filter frequency response');
```

Note: The frequency response of a filter can be also visualized with the fvtool GUI.

Finally, let us present a very simple example of a speech denoising technique, which goes through the following steps:

1. Loads a speech signal from a WAVE file.
2. Generates a white noise signal of equal duration with the speech signal.
3. Uses a highpass filter to preserve the frequencies of the noise above 3 kHz.
4. Adds the filtered noise to the original signal.
5. Denoises the signal with a lowpass filter.
6. Reproduces all three signals and plots their spectrograms.

The MATLAB code that implements this simple denoising approach is the following (stored in the demoFilter2() m-file of the library):

```
[x,Fs] = wavread(['..' filesep 'data' ...
         filesep 'diarizationExample.wav'], [280000 300000]);
x = x / max(abs(x));                        % normalize
N = rand(size(x)); N = N / max(abs(N));     % create white noise
Fst = 3000; Fp  = 3050; Ap = 0.5; As = 30;  % highpass — filter the noise:
hpFilter = fdesign.highpass('Fst,Fp,Ast,Ap', Fst, Fp, As, Ap, Fs);
F = design(hpFilter,'equiripple'); b = F.Numerator; N = filter(b, 1, N);
% add noise to signal (weighted):
w = 0.05; xN = w * x + (1—w) * N; xN = xN / max(abs(xN));
% low—pass filter the noisy signal
Fpass = 3000; Fstop = 3050; Ap = 0.5; As = 30;
lpFilter = fdesign.lowpass('Fp,Fst,Ap,Ast', Fpass , Fstop, Ap, As, Fs);
F = design(lpFilter,'equiripple'); b = F.Numerator;
y = filter(b, 1, xN); y = y / max(abs(y));
figure; subplot(2, 2, 1);                   % plot and playback results
spectrogram(x, 0.020 * Fs, 0.010 * Fs, Fs/2, Fs); axis xy;   view(90,90);
title('Spectrogram — original signal'); subplot(2, 2, 2);
spectrogram(N, 0.020 * Fs, 0.010 * Fs, Fs/2, Fs); axis xy;   view(90,90);
title('Spectrogram — noise'); subplot(2, 2, 3);
spectrogram(xN, 0.020 * Fs, 0.010 * Fs, Fs/2, Fs); axis xy;   view(90,90);
title('Spectrogram — noisy signal'); subplot(2, 2, 4);
spectrogram(y, 0.020 * Fs, 0.010 * Fs, Fs/2, Fs); axis xy;   view(90,90);
title('Spectrogram — filtered (deNd) signal');
fprintf('Playing original signal...\n'); sound(x, Fs);
fprintf('Playing noisy signal...\n'); sound(xN, Fs);
fprintf('Playing filtered signal...\n'); sound(y, Fs);
```

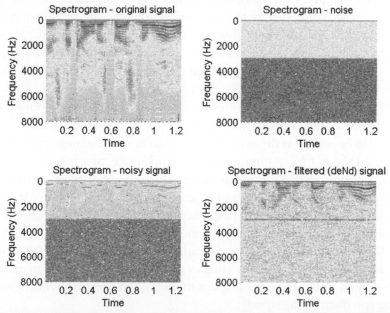

Figure 3.9 Example of a simple speech denoising technique applied on a segment of the diarizationExample.wav file, found in the data folder of the library of the book. The figure represents the spectrograms of the original signal, the high-frequency noise, the noisy signal, and the final filtered signal.

During the playback of the noisy signal, the speech signal can be hardly perceived, whereas this is not the case with the filtered signal. However, note that this demo is only presenting a naive denoising procedure. In a real-world scenario the noise is not expected to be limited to high frequencies and therefore more sophisticated denoising approaches need to be adopted. Figure 3.9 illustrates the results of this filtering process for a speech signal example.

3.8. EXERCISES

1. (D3) Use the `fdesign` object to design a bandpass filter that attenuates all frequencies outside the range [100–500]Hz. Apply it on audio signals of your choice, plot the spectrograms of the results, and reproduce the respective sounds.

2. (D5) Use the `GUIDE` tool of MATLAB to create a GUI that implements the following requirements:
   (a) When a button is pressed, the GUI records 2 s of audio (see also Section 2.6), using MATLAB's `audiorecorder`, as it was explained in Section 2.6.2. The recorded signal is stored in a temporary WAV file.
   (b) Generates a FIR bandpass filter using the `fdesign` tool. The frequencies that specify the passband of the filter are provided by the user by means of two sliders.
   (c) Uses a variant of the `stpFile()` function (see also Section 2.7) to apply the bandpass filter on each short-term window of the recorded signal.
   (d) When a 'play' button is pressed, the filtered signal is reproduced.

3. (D1) We are given the signal $x(n)$, $n = 0, 1, \ldots, 999$. How many zeros need to be padded at the end of the signal, so that the frequency resolution of the DFT is 4 Hz, assuming that the sampling frequency is 16, 000 Hz?

4. (D3) Assume that we want to design a pre-emphasis filter, so that the magnitude of its frequency response is equal to 0.739 on a linear scale. Assuming that the sampling frequency is 8000 Hz, what value should we use for the a parameter in the equation $y(n) = x(n) + a \cdot x(n-1)$?

5. (D2) The analog signal $x_\alpha(t) = 0.3cos(2\pi 300\pi t) + 0.2cos(2\pi 310t)$ is sampled at 16 kHz. How many samples are necessary so that the frequency resolution of the DFT is sufficient to distinguish the two frequencies in the spectrum of the signal?

6. (D2) The analog signal $x_\alpha(t) = 0.1cos(2\pi 100\pi t) + 0.3cos(2\pi 1100t) + 0.1cos(2\pi 2100\pi t)$ is sampled with a 1 kHz sampling frequency. What is the resulting discrete-time signal?

7. (D3) Let $x(n), n = 0, \ldots, N - 1$ be a signal and $y(n), n = 0, \ldots, N - 1$ its reconstruction. The difference between the original signal and its reconstruction can be quantified by the signal-to-noise ratio (SNR), which is defined as:

$$\text{SNR} = 10 \log_{10} \frac{\sum_{n=0}^{N-1} x^2(n)}{\sum_{n=0}^{N-1} (x(n) - y(n))^2}.$$

Implement a function that computes the SNR given a signal and its reconstruction. Compute the SNR of the DCT compression scheme (Section 3.4) and DTWT reconstruction procedure. Use the speech segment of the `diarizationExample.wav` file, as in Section 3.5. You can also experiment with different values of the compression ratio to investigate how this parameter affects the respective SNR.

Audio Features

Contents

Feature extraction is an important audio analysis stage. In general, feature extraction is an essential processing step in pattern recognition and machine learning tasks. The goal is to extract a set of features from the dataset of interest. These features must be informative with respect to the desired properties of the original data. Feature extraction can also be viewed as a data rate reduction procedure because we want our analysis algorithms to be based on a relatively small number of features. In our case, the original data, i.e. the audio signal, is voluminous and as such, it is hard to process directly in any analysis task. We therefore need to transform the initial data representation to a more suitable one, by extracting audio features that represent the properties of the original signals while reducing the volume of data. In order to achieve this goal, it is important to have a good knowledge of the application domain, so that we can decide which are the best features. For example, when discriminating between speech and music segments, an interesting feature candidate is the deviation of the signal's energy, because this feature has a physical meaning that fits well with the particular classification task (a more detailed explanation will be given in a later section).

In this chapter, we present some essential features which have been widely adopted by several audio analysis methods. We describe how to extract these features on a short-term basis and how to compute feature statistics for the case of mid-term audio segments. The adopted short-term features and mid-term statistics will be employed throughout this book by several audio analysis techniques such as audio segment classification and audio segmentation. Our purpose is not to present every audio feature that has been proposed in the research literature, as this would require several chapters; instead, using a MATLAB programming approach, we wish to provide reproducible descriptions of the key ideas which underlie audio feature extraction.

4.1. SHORT-TERM AND MID-TERM PROCESSING

4.1.1. Short-Term Feature Extraction

As explained in Section 2.7, in most audio analysis and processing methods, the signal is first divided into short-term frames (windows). This approach is also employed during the feature extraction stage; the audio signal is broken into possibly overlapping frames and a set of features is computed per frame. This type of processing generates a sequence, $\mathbf{F}$, of feature vectors per audio signal. The dimensionality of the feature vector depends on the nature of the adopted features. It is not uncommon to use one-dimensional features, like the energy of a signal, however, in most sophisticated audio analysis applications several features are extracted and combined to form feature vectors of increased dimensionality. The extracted sequence(s) of feature vectors can then be used for subsequent processing/analysis of the audio data.

In this line of thinking, function stFeatureExtraction() generates, on a short-term processing basis, 23 features given an audio signal.

```
function Features = stFeatureExtraction(signal, fs, win, step)

% function Features = stFeatureExtraction(signal, fs, win, step)
%
% This function computes basic audio feature sequencies for an audio
% signal, on a short—term basis.
%
% ARGUMENTS:
% — signal:    the audio signal
% — fs:        the sampling frequency
```

```
%  — win:        short—term window size (in seconds)
%  — step:       short—term step (in seconds)
%
% RETURNS:
%  — Features: a [MxN] matrix, where M is the number of features and N is
%  the total number of short—term windows. Each line of the matrix
%  corresponds to a seperate feature sequence
%

% if STEREO ...
if (size(signal,2)>1)
    signal = (sum(signal,2)/2); % convert to MONO
end

% convert window length and step from seconds to samples:
windowLength = round(win * fs);
step = round(step * fs);

curPos = 1;
L = length(signal);

% compute the total number of frames:
numOfFrames = floor((L—windowLength)/step) + 1;
% number of features to be computed:
numOfFeatures = 35;
Features = zeros(numOfFeatures, numOfFrames);
Ham = window(@hamming, windowLength);
mfccParams = feature_mfccs_init(windowLength, fs);
for i=1:numOfFrames % for each frame
    % get current frame:
    frame  = signal(curPos:curPos+windowLength—1);
    frame  = frame .* Ham;
    frameFFT = getDFT(frame, fs);
    if (sum(abs(frame))>eps)
        % compute time—domain features:
        Features(1,i) = feature_zcr(frame);
        Features(2,i) = feature_energy(frame);
        Features(3,i) = feature_energy_entropy(frame, 10);

        % compute freq—domain features:
        if (i==1) frameFFTPrev = frameFFT; end;
        [Features(4,i) Features(5,i)] = ...
            feature_spectral_centroid(frameFFT, fs);
        Features(6,i) = feature_spectral_entropy(frameFFT, 10);
        Features(7,i) = feature_spectral_flux(frameFFT, frameFFTPrev);
        Features(8,i) = feature_spectral_rolloff(frameFFT, 0.90);
        MFCCs = feature_mfccs(frameFFT, mfccParams);
        Features(9:21,i)  = MFCCs;

        [HR, F0] = feature_harmonic(frame, fs);
        Features(22, i) = HR;
```

```
        Features(23, i) = F0;

        Features(23+1:23+12, i) = feature_chroma_vector(frame, fs);

    else

        Features(:,i) = zeros(numOfFeatures, 1);

    end

    curPos = curPos + step;

    frameFFTPrev = frameFFT;

end

Features(35, :) = medfilt1(Features(35, :), 3);
```

The above function works by breaking the audio input into short-term windows and computing 23 audio features per window. Lines 48–65 call the respective feature extraction functions (which will be described in the next sections). For instance, function `feature_zcr()` computes the zero-crossing rate of an audio frame.

4.1.2. Mid-Term Windowing in Audio Feature Extraction

Another common technique is the processing of the feature sequence on a mid-term basis. According to this type of processing, the audio signal is first divided into mid-term segments (windows) and then, *for each segment,* the short-term processing stage is carried out.[1] At a next step, the feature sequence, **F**, which has been extracted from a mid-term segment, is used for computing feature statistics, e.g. the average value of the zero-crossing rate. In the end, *each mid-term segment is represented by a set of statistics* which correspond to the respective short-term feature sequences. During mid-term processing, we assume that the mid-term segments exhibit homogeneous behavior with respect to audio type and it therefore makes sense to proceed with the extraction of statistics on a segment basis. In practice, the duration of mid-term windows typically lies in the range 1–10 s, depending on the application domain. Figure 4.1 presents the process of extracting mid-term statistics of audio features.

The next function takes as input short-term feature sequences that have been generated by `stFeatureExtraction()` and returns a vector that contains the resulting feature statistics. If, for example, 23 feature sequences have been computed on a short-term basis and two mid-term statistics are drawn per feature (e.g. the mean value and the standard deviation of the feature), then, the output of the mid-term function is a 46-dimensional vector. The structure of this vector is the following: elements 1 and 24

---

[1] In the literature, mid-term segments are sometimes referred to as "texture" windows.

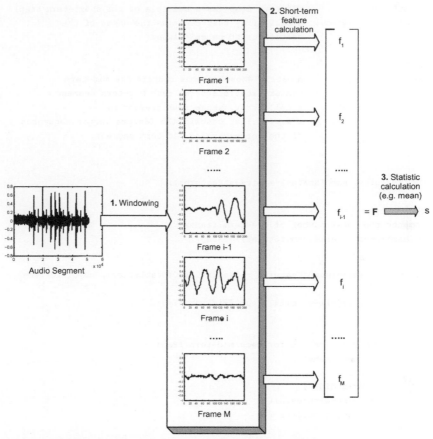

Figure 4.1 Mid-term feature extraction: each mid-term segment is short-term processed and statistics are computed based on the extracted feature sequence.

correspond to the mean and standard deviation of the first short-term feature sequence, elements 2 and 25 correspond to the mean and standard deviation of the second audio short-term sequence, and so on.

```
function [mtFeatures, shortFeaturesCell] = ...
    mtFeatureExtraction(stFeatures, mtWin, mtStep, listOfStatistics)


%
% This function is used for extracting mid—term statistics
%
% ARGUMENTS:
%  — stFeatures:        a matrix that contains all short—term feature vectors
%                       (dimension: dFeatures x numOfShortTermWindows)
%  — mtWin:             mid—term window (as a multiple of short—term window)
```

```
%  — mtSteP:               mid—term step (as a multiple of the short—term step)
%  — listOfStatistics:     a cell array that contains the names of the
%                          statistics to be calculated
%
% RETURNS:
%  — mtFeatures:           a matrix whose columns contain the mid—term
%                          feature statistics for each mid—term segment
%  — stFeaturesCell:       a cell array, whose i—th element is
%                          a matrix that contain the feature vector sequences
%                          of the corresponding mid—term segment.
%
%

[numOfFeatures, numOfStWins] = size(stFeatures);

curPos = 1;
% compute the total number of mid—term frames:
numOfMidFrames = ceil((numOfStWins)/mtStep);


mtFeatures = zeros(numOfFeatures * length(listOfStatistics), numOfMidFrames);
if (nargout==2)
    shortFeaturesCell = cell(1, numOfMidFrames);
end

for (i=1:numOfMidFrames) % for each mid—term frame
    % get current frame:
    N1 = curPos;
    N2 = curPos+mtWin—1;
    if (N2>size(stFeatures,2))
        N2 = size(stFeatures,2);
    end

        CurStFeatures  = stFeatures(:, N1:N2);
        if (nargout==2)
            shortFeaturesCell{i} = CurStFeatures;
        end
        for (j=1:length(listOfStatistics))
            mtFeatures( (j—1)*numOfFeatures + 1: j*numOfFeatures, i) = ...

                computeStatistic(CurStFeatures', listOfStatistics{j});
        end
    curPos = curPos + mtStep;
end


function S = computeStatistic(seq, statistic)
    if strcmpi(statistic, 'mean')
```

```
      S = mean(seq); return;
end
if strcmpi(statistic, 'median')
      S = median(seq); return;
end
if strcmpi(statistic, 'std')
      S = std(seq); return;
end
if strcmpi(statistic, 'stdbymean')
      S = std(seq) ./ (mean(seq)+eps); return;
end
if strcmpi(statistic, 'max')
      S = max(seq); return;
end
if strcmpi(statistic, 'min')
      S = min(seq); return;
end
if strcmpi(statistic, 'meanNonZero')
      for i=1:size(seq, 2)
          curSeq = seq(:, i);
          S(i) = mean(curSeq(curSeq>0));
      end
      return;
end
if strcmpi(statistic, 'medianNonZero')
      for i=1:size(seq, 2)
          curSeq = seq(:, i);
          S(i) = median(curSeq(curSeq>0));
      end
      return;
end
```

Note: In some cases, the mid-term feature extraction process can be employed in a longer time-scale scenario, in order to capture salient features of the audio signal. For example, in the context of genre classification of music tracks [14], it is often desirable to extract a single feature vector as the representative of the whole music signal. In such cases, short-term features are first extracted as described above and at an intermediary step, mid-term statistics are computed on a mid-term segment basis (e.g. every 2 s). At a final stage, the mid-term statistics are *long-term averaged*, in order to provide a single vector representation of the whole signal. It is worth noting that the final long-term averaging operation highlights the outstanding (salient) features of the music signal but discards details related to its temporal evolution. However, this long-term averaging approach has exhibited acceptable performance over the years for the task of music genre classification [14] and related problems.

4.1.3. Extracting Features from an Audio File

As explained above, function `stFeatureExtraction()` generates short-term feature sequences from an audio signal, and function `mtFeatureExtraction()` extracts sequences of mid-term statistics based on the previously extracted short-term features. We will now show how to break a large audio file (or audio stream) into mid-term windows and generate the respective mid-term audio statistics. As shown in Section 2.5, if the audio file to be analyzed has a very long duration, then loading all its content in one step can be prohibitive due to memory issues. This is why function `readWavFile()` demonstrated how to read progressively the contents of a large audio file using chunks of data (blocks), which can be, for example, one minute long. In the same line of thinking, function `featureExtractionFile()` is also using data blocks to analyze large audio files. The input arguments of this function are:

- The name of the audio file to be analyzed.
- The short-term window size (in seconds).
- The short-term window step (in seconds).
- The mid-term window size (in seconds).
- The mid-term window step (in seconds).
- A cell array with the names of the feature statistics which will be computed on a mid-term basis.

  The arguments at the output of `featureExtractionFile()` are:

- A matrix, whose rows contain the mid-term sequences of audio statistics. For example, the first row may correspond to the mean value of short-term energy (all features will be explained later in this chapter).
- The time instant that marks the center of each mid-term segment (in seconds).
- A cell array, each element of which contains the short-term feature sequences of the respective mid-term segment.

  The source code of function `featureExtractionFile()` is available on the Audio Analysis Library that accompanies the book. We now demonstrate the use of `featureExtractionFile()` via the example introduced by function `plotFeaturesFile()`:

```
function plotFeaturesFile(wavFileName, featureToPlot)

%
% function plotFeaturesFile(wavFileName, featureToPlot)
%
% This function is used to plot feature sequences and respective
% mid—term statistics
%
% Example:
% plotFeaturesFile('diarizationExample.wav',6)

% feature extraction parameters (windows and statistics):
shortTermSize = 0.050; shortTermStep = 0.025;
midTermSize = 2.0; midTermStep = 1.0;
Statistics = {'mean','median','std','stdbymean','max','min'};
% feature extraction (mid—term and short—term):
[midFeatures, Centers, stFeaturesPerSegment] = featureExtractionFile(...
    wavFileName, shortTermSize, shortTermStep, midTermSize, midTermStep, Statistics);
numOfShortFeatures = size(stFeaturesPerSegment{1}, 1);
% Plot results:
figure; hold on; Colors = {'r', 'g', 'k', 'm', 'y','c'};
% Plot mid—term feature statistics:
for s=1:length(Statistics) % for each statistic:
    P = plot(Centers, ...
        midFeatures(featureToPlot + numOfShortFeatures * (s—1), :), Colors{s});
    set(P, 'linewidth', 2);
end
% Plot short—term feature sequence:
for i=1:length(stFeaturesPerSegment) % for each mid—term window:
    % get current short—term feature sequence to be plotted:
    curSFeature = stFeaturesPerSegment{i}(featureToPlot, :);
    % create time array:
    stTime = (i—1) * midTermStep : shortTermStep : ...
        (i—1) * midTermStep + (length(curSFeature)—1)*shortTermStep;
    % plot the respective short—term feature sequence:
    plot(stTime, curSFeature);
end
legend([Statistics, 'short—term sequence']);
xlabel('Time (seconds)'); ylabel('Feature Values');
title(['Feature ' num2str(featureToPlot)]);
```

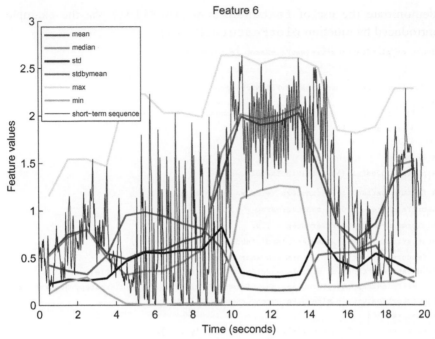

Figure 4.2 Plotting the results of `featureExtractionFile()`, using `plotFeaturesFile()`, for the six feature statistics drawn from the 6th adopted audio feature.

For example, type:

```
plotFeaturesFile('music4genres', 6)
```

The result will be that the short-term feature sequence of the 6th audio feature and all respective mid-term statistics will be plotted. More specifically, further to the 6th short-term feature sequence, function `plotFeatures File()` will also plot the 6th, $6+D$th, ..., $6+(S-1)D$th statistics, where D is the total number of features and S is the number of adopted statistics. In other words, it will plot all the statistics that are relevant to the 6th short-term feature. In this case, $S = 6$ and $D = 23$, therefore, the dimensionality of the final mid-term feature vector is 138. The outcome of the above call to `plotFeaturesFile()` is presented in Figure 4.2.

4.2. CLASS DEFINITIONS

For the sake of experimentation and as a means to demonstrate the properties of the adopted features and corresponding statistics, we now proceed with two audio analysis examples. The first experimental setup employs audio segments from eight audio classes which are frequently encountered when processing the audio content of movies and radio broadcasts. The second example focuses on music genre classification. Before we proceed with the theoretical and practical descriptions of the audio features, let us briefly present the audio classes that have been involved in our two examples.

The first set of audio classes refers to the problem of detecting/tracking audio events in movies. In the context of content-based processing, the detection of audio events can play an important role in the development of solutions that can deal with important problems, like violence detection in movies and video streams from video sharing sites. In our case, the class names and respective descriptions are summarized in Table 4.1.

In the second example, we use a set of music tracks to demonstrate the discriminative capability of selected features for the problem of *musical genre classification*. For pedagogical purposes, we use 360 tracks belonging to three music genres, namely classical, jazz, and electronic (120 samples per class). The reader may also find it interesting to experiment with a popular dataset for music genre classification, i.e. the GTZAN dataset [15], on which several music genre classification algorithms have been tested over the past 10 years.

Table 4.1 Class Descriptions for the Multi-Class Task of Movie Segments

Class ID	Class Name	Class Description
1	Music	Music from film soundtracks and music effects
2	Speech	Speech segments from various speakers, languages, and emotional states. These segments exhibit diverse levels of "noisiness" because speech is usually mixed with other types of audio classes (especially in films soundtracks)
3	Others1	Environmental sounds of low energy and practically stable signal intensity (e.g. silence, background noise, wind, rain, etc.)
4	Others2	Environmental sounds with abrupt changes in signal energy (e.g. a door closing, the sound of thunder, an object breaking, etc.)
5	Others3	High energy, non–abrupt (quasi-repeating) environmental sounds (e.g. machinery)
6	Gunshots	Gunshots from various gun types. This class contains short, abrupt sounds, and continuous (repetitive) gunshots
7	Fights	The sounds of humans fighting
8	Screams	Human screams

> *Note:* The two classification tasks involved in the current chapter are just representative; other classification tasks have also been trained and evaluated, especially in Chapter 5, in the context of presenting several fundamental classification concepts. However, the audio classes covered by these two tasks are enough to demonstrate the discrimination ability of audio features.

We now start with the presentation of some of the most important and widely used audio features, along with the corresponding MATLAB code and examples. We also discuss a number of statistics, which, on a mid-term basis, provide acceptable discrimination capability among the audio classes, in the context of the presented examples. Note that we assume that the input to the feature extraction functions is an audio frame and therefore, the functions are to be called inside a short-term analysis process, as it is the case with function `stFeatureExtraction()`. Similarly, the computation of the mid-term statistics can be encapsulated in functions like `mtFeatureExtraction()` that operate on consecutive audio segments.

4.3. TIME-DOMAIN AUDIO FEATURES

In general, the time-domain audio features are extracted directly from the samples of the audio signal. Typical examples are the short-term energy and short-term zero-crossing rate. Such features offer a simple way to analyze audio signals, although it is usually necessary to combine them with more sophisticated frequency-domain features, which will be described in Section 4.4. What follows, are definitions of some of the most celebrated time-domain features.

4.3.1. Energy

Let $x_i(n)$, $n = 1, \ldots, W_L$ be the sequence of audio samples of the ith frame, where W_L is the length of the frame. The short-term energy is computed according to the equation

$$E(i) = \sum_{n=1}^{W_L} |x_i(n)|^2. \tag{4.1}$$

Usually, energy is normalized by dividing it with W_L to remove the dependency on the frame length. Therefore, Eq. (4.1) becomes

$$E(i) = \frac{1}{W_L} \sum_{n=1}^{W_L} |x_i(n)|^2.$$

(4.2)

Equation (4.2) provides the so-called *power* of the signal. In the rest of this chapter, we will use the power of the signal in our feature extraction functions but for the sake of simplicity we will keep using the term 'energy' in the respective algorithmic descriptions.

The following function extracts the energy value of a given audio frame:

```
function E = feature_energy(window)

% function E = feature_energy(window);
%
% This function calculates the energy of an audio frame.
%
% ARGUMENTS:
% − window:      an array that contains the audio samples of the input frame
%
% RETURN:
% − E:      the computed energy value
%
E = (1/(length(window))) * sum(abs(window.^2));
```

Short-term energy is expected to exhibit high variation over successive speech frames, i.e. the energy envelope is expected to alternate rapidly between high and low energy states. This can be explained by the fact that speech signals contain weak phonemes and short periods of silence between words. Therefore, a mid-term statistic that can be used for classification purposes in conjunction with short-term energy is the standard deviation σ^2 of the energy envelope. An alternative statistic for short-term energy, which is independent of the intensity of the signal, is the standard deviation by mean value ratio, ($\frac{\sigma^2}{\mu}$) [16]. Figure 4.3 presents histograms of the standard deviation by mean ratio for segments of the classes: music and speech. It has been assumed that each segment is homogeneous, i.e. it contains either music or speech. The figure indicates that the values of this statistic are indeed higher for the speech class. The Bayesian error for the respective binary classification task was found to be equal to 17.8% for this experiment, assuming that the two classes are a priori equiprobable and that the likelihood of the statistic (feature) given the class is well approximated using the extracted histograms. Note, that the Bayesian error for the same binary classification task (music vs speech), when the standard deviation statistic is used instead, increases

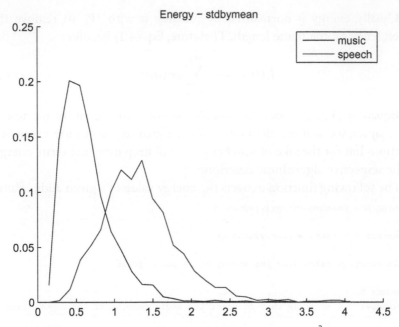

Figure 4.3 Histograms of the standard deviation by mean ratio $(\frac{\sigma^2}{\mu})$ of the short-term energy for two classes: music and speech. Speech segments favor higher values of this statistic. The Bayesian error for the respective binary classification task was 17.8%.

to 37%. Therefore, the normalization of the standard deviation feature by dividing it with the respective mean value provides a crucial performance improvement in this case.

Note: The term 'Bayesian error' that we have adopted in this chapter will be further described in Chapter 5, where we introduce several fundamental classification concepts. For the moment, it suffices to say that the Bayesian error refers to the classification performance of the Bayesian classifier, which bases its classification decisions on posterior class probabilities and is known to be *optimal* with respect to the probability of error. Therefore, the Bayesian classifier provides a *performance bound* for every classifier. In this chapter we use the term Bayesian error in a looser sense, i.e. we assume that the classes are equiprobable and that the feature histograms can be used to approximate successfully the probability density function of a feature given the class, thanks to the large amount of audio samples that are used to generate the histograms. As will become evident in Chapter 5, these assumptions permit us to imitate the operation of the Bayesian classifier via an approximation, although strictly speaking, we are not using the Bayesian classifier per se.

4.3.2. Zero-Crossing Rate

The Zero-Crossing Rate (ZCR) of an audio frame is the rate of sign-changes of the signal during the frame. In other words, it is the number of times the signal changes value, from positive to negative and vice versa, divided by the length of the frame. The ZCR is defined according to the following equation:

$$Z(i) = \frac{1}{2W_L} \sum_{n=1}^{W_L} | \, sgn[x_i(n)] - sgn[x_i(n-1)] \, |, \qquad (4.3)$$

where $sgn(\cdot)$ is the sign function, i.e.

$$sgn[x_i(n)] = \begin{cases} 1, & x_i(n) \geq 0, \\ -1, & x_i(n) < 0. \end{cases} \qquad (4.4)$$

In our toolbox, the computation of the zero-crossing rate for a given frame is implemented in the following m-file:

```
function Z = feature_zcr(window);

% function  Z = feature_zcr(window);
%
% This function calculates the zero crossing rate of an audio frame.
%
% ARGUMENTS:
% - window:    an array that contains the audio samples of the input frame
%
% RETURN:
% - Z:         the computed zero crossing rate value
%

window2 = zeros(size(window));
window2(2:end) = window(1:end-1);
Z = (1/(2*length(window))) * sum(abs(sign(window)-sign(window2)));
```

ZCR can be interpreted as a measure of the noisiness of a signal. For example, it usually exhibits higher values in the case of noisy signals. It is also known to reflect, in a rather coarse manner, the spectral characteristics of a signal [17]. Such properties of the ZCR, along with the fact that it is

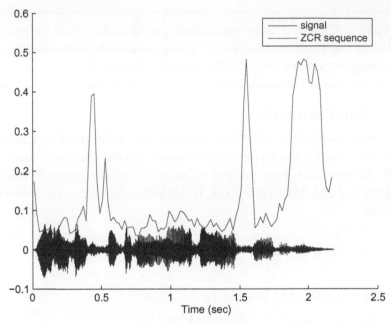

Figure 4.4 Example of a speech segment and the respective sequence of ZCR values.

easy to compute, have led to its adoption by numerous applications, including speech–music discrimination [18,16], speech detection [19], and music genre classification [14], to name but a few.

Figure 4.4 presents a speech signal along with the respective ZCR sequence. It shows that the values of ZCR are higher for the noisy parts of the signal, while in speech frames the respective ZCR values are generally lower (depending, of course, on the nature and context of the phoneme that is pronounced each time).

An interesting observation is that the standard deviation of this feature over successive frames is higher for speech signals than for music signal. Indeed, in Figure 4.5 we present the histograms of the standard deviation of ZCR over the short-term frames of music and speech segments. For the generation of these figures, more than 400 segments were used per class. It can be seen that the values of standard deviation are much higher for the speech class. So, in this case, if the standard deviation of the ZCR is used as a mid-term feature for the binary classification task of speech vs music, the respective Bayesian error [5] is equal to 22.3%. If the mean value of the ZCR is used instead, the classification error increases to 34.2%, for this example.

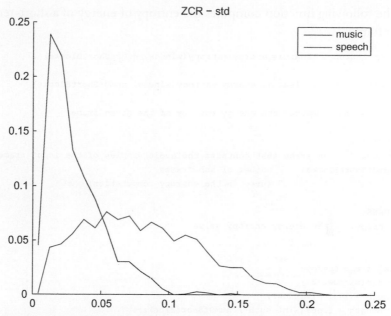

Figure 4.5 Histograms of the standard deviation of the ZCR for music and speech classes. Speech segments yield higher feature values. The respective Bayesian error for this binary classification task is 22.3%.

4.3.3. Entropy of Energy

The short-term entropy of energy can be interpreted as a measure of abrupt changes in the energy level of an audio signal. In order to compute it, we first divide each short-term frame in K sub-frames of fixed duration. Then, for each sub-frame, j, we compute its energy as in Eq. (4.1) and divide it by the total energy, $E_{shortFrame_i}$, of the short-term frame. The division operation is a standard procedure and serves as the means to treat the resulting sequence of sub-frame energy values, $e_j, j = 1, \ldots, K$, as a sequence of probabilities, as in Eq. (4.5):

$$e_j = \frac{E_{subFrame_j}}{E_{shortFrame_i}}, \tag{4.5}$$

where

$$E_{shortFrame_i} = \sum_{k=1}^{K} E_{subFrame_k}. \tag{4.6}$$

At a final step, the entropy, $H(i)$ of the sequence e_j is computed according to the equation:

$$H(i) = -\sum_{j=1}^{K} e_j \cdot \log_2 (e_j). \tag{4.7}$$

The following function computes the entropy of energy of a short-term audio frame:

```
function Entropy = feature_energy_entropy(window, numOfShortBlocks)

% function Entropy = feature_energy_entropy(window, numOfShortBlocks)
%
% This function computes the energy entropy of the given frame
%
% ARGUMENTS:
% — window:       an array that contains the audio samples of the input frame
% — numOfShortBlocks:     number of sub-frames
%                         (used in the entropy computation)
%
% RETURNS:
% — Entropy:    the energy entropy value
%

% total frame energy:
Eol = sum(window.^2);
winLength = length(window);
subWinLength = floor(winLength / numOfShortBlocks);

if length(window)≠subWinLength * numOfShortBlocks
    window = window(1:subWinLength * numOfShortBlocks);
end
% get sub-windows:
subWindows = reshape(window, subWinLength, numOfShortBlocks);

% compute normalized sub-frame energies:
s = sum(subWindows.^2) / (Eol+eps);

% compute entropy of the normalized sub-frame energies:
Entropy = —sum(s.*log2(s+eps));
```

The resulting value is lower if abrupt changes in the energy envelope of the frame exist. This is because, if a sub-frame yields a high energy value, then one of the resulting probabilities will be high, which in turn reduces the entropy of sequence e_j. Therefore, this feature can be used for the detection of significant energy changes, as it is, for example, the case with the beginning of gunshots, explosions, various environmental sounds, etc. In Figure 4.6, an example of the sequence of entropy values is presented for an audio signal that contains three gunshots. It can be seen that in the beginning of each gunshot the value of this feature decreases. Several research efforts have

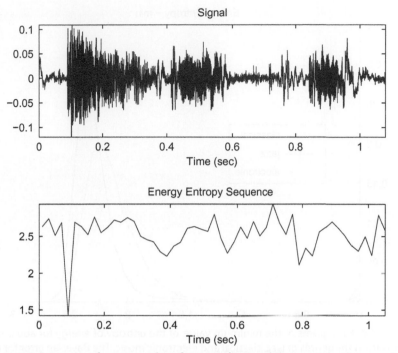

Figure 4.6 Sequence of entropy values for an audio signal that contains the sounds of three gunshots. Low values appear at the onset of each gunshot.

experimented with the entropy of energy in the context of detecting the onset of abrupt sounds, e.g. [20, 21].

Figure 4.7 presents a second example, which is based on computing a long-term statistic of this feature for segments belonging to the three musical genres that were previously examined. Specifically, for each music segment, we have selected the minimum of the sequence of entropy values as the long-term feature that will eventually be used to discriminate among genres. It can be seen that this long-term feature takes lower values for electronic music and higher values for classical music (although a similar conclusion cannot be reached for Jazz). This can be partly explained by the fact that electronic music tends to contain many abrupt energy changes (low entropy), compared to classical music, which exhibits a smoother energy profile. The reader is reminded that such examples serve pedagogical needs and do not imply that the adopted features are optimal in any sense or that they can necessarily lead to acceptable performance in a real-world scenario.

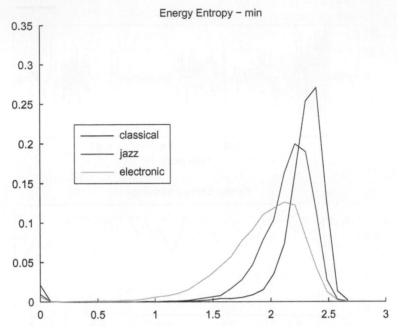

Figure 4.7 Histograms of the minimum value of the entropy of energy for audio segments from the genres of jazz, classical and electronic music. The Bayesian error for this three-class classification task, using just this single feature, was equal to 44.8%.

4.4. FREQUENCY-DOMAIN AUDIO FEATURES

As discussed in Section 3.1, the Discrete Fourier Transform (DFT) of a signal can be easily computed with MATLAB using the built-in function `fft()`. DFT is widely used in audio signal analysis because it provides a convenient representation of the distribution of the frequency content of sounds, i.e. of the sound spectrum. We will now describe some widely used audio features that are based on the DFT of the audio signal. Features of this type are also called frequency-domain (or spectral) audio features.

In order to compute the spectral features, we must first compute the DFT of the audio frames using the `getDFT()` function (Section 3.1). Function `stFeatureExtraction()`, which has already been described in 4.1, computes the DFT of each audio frame (calling function `getDFT()`), and the resulting DFT coefficients are used to compute various spectral features. In order to proceed, let $X_i(k)$, $k = 1, \ldots, W_{FL}$, be the magnitude of the DFT coefficients of the ith audio frame (at the output of function `getDFT()`).

In the following paragraphs, we will describe how the respective spectral features are computed based on the DFT coefficients of the audio frame.

Note: W_L is the number of samples per short-term window (frame). This is also the number of DFT coefficients of the frame. Roughly speaking, for the computation of spectral features, it suffices to work with the first half of the coefficients, because the second half mainly serves to reconstruct the original signal. For notational simplicity, let Wf_L *be the number of coefficients that are used in the computations to follow.*

4.4.1. Spectral Centroid and Spread

The spectral centroid and the spectral spread are two simple measures of spectral position and shape. The spectral centroid is the center of 'gravity' of the spectrum. The value of spectral centroid, C_i, of the ith audio frame is defined as:

$$C_i = \frac{\sum_{k=1}^{Wf_L} kX_i(k)}{\sum_{k=1}^{Wf_L} X_i(k)}. \tag{4.8}$$

Spectral spread is the second central moment of the spectrum. In order to compute it, one has to take the deviation of the spectrum from the spectral centroid, according to the following equation:

$$S_i = \sqrt{\frac{\sum_{k=1}^{Wf_L} (k - C_i)^2 X_i(k)}{\sum_{k=1}^{Wf_L} X_i(k)}}. \tag{4.9}$$

The MATLAB code that computes the spectral centroid and spectral spread of an audio frame is presented in the following function. Note that this function (like all spectral-based functions of this chapter) takes as input the magnitude of the DFT coefficients of an audio frame (output of the `getDFT()` function), instead of the audio frame itself. The reader is prompted to revisit Section 4.1 and in particular the `stFeature Extraction()` function, in order to study how the DFT and related spectral features are computed on a short-term basis. Finally, we normalize both features in the range $[0,1]$, by dividing their values by $\frac{F_s}{2}$. This type of normalization can be very useful when features that take frequency values are combined with other features in audio analysis tasks.

```
function [C,S] = feature_spectral_centroid(window_FFT, fs)

% function C = feature_spectral_centroid(window_FFT, fs)
%
% Computes the spectral centroid and spread of a frame
%
% ARGUMENTS:
% — window_FFT: the abs(FFT) of an audio frame
%               (computed by getDFT() function)
% — fs:         the sampling freq of the input signal (in Hz)
%
% RETURNS:
% — C:          the value of the spectral centroid
%               (normalized in the 0..1 range)
% — S:          the value of the spectral spread
%               (normalized in the 0..1 range)
%

% number of DFT coefficients:
windowLength = length(window_FFT);
% sample range
m = ((fs/(2*windowLength))*[1:windowLength])';

% normalize the DFT coefs by the max value:
window_FFT = window_FFT / max(window_FFT);
% compute the spectral centroid:
C = sum(m.*window_FFT)/ (sum(window_FFT)+eps);
% compute the spectral spread
S = sqrt(sum(((m-C).^2).*window_FFT)/ (sum(window_FFT)+eps));

% normalize by fs/2
% (so that 1 correponds to the maximum signal frequency, i.e. fs/2):
C = C / (fs/2);
S = S / (fs/2);
```

Returning to the spectral centroid, it can be observed that higher values correspond to brighter sounds. In Figure 4.8, we present the histograms of the maximum value of the spectral centroid for audio segments from three classes of environmental sounds. It can be seen that the class 'others1,' which mostly consists of background noise, silence, etc. exhibits lower values for this statistic, while the respective values are higher for the abrupt sounds of classes "others2" and "others3".

On the other hand, spectral spread measures how the spectrum is distributed around its centroid. Obviously, low values of the spectral spread correspond to signals whose spectrum is tightly concentrated around the spectral centroid. In Figure 4.9, we present the histograms of the maximum value (mid–term statistic) of the sequence of values of the spectral spread feature for audio segments that stem from three musical classes (classical, jazz, and

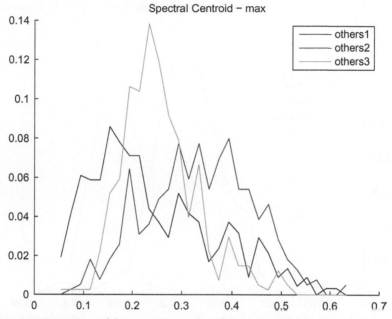

Figure 4.8 Histograms of the maximum value of the sequence of values of the spectral centroid, for audio segments from three classes of environmental sounds: others1, others2, and others3. The Bayesian error for this three-class task, using this single feature, was equal to 44.6%.

electronic). In this specific example, the results indicate that the spectrograms of excerpts of electronic music are (usually) more widely spread around their centroid than (especially) classical and jazz music. It should not be taken for granted that this observation is the same for every music track in these genres.

4.4.2. Spectral Entropy

Spectral entropy [22] is computed in a similar manner to the entropy of energy, although, this time, the computation takes place in the frequency domain. More specifically, we first divide the spectrum of the short–term frame into L sub-bands (bins). The energy E_f of the fth sub-band, $f = 0, \ldots, L - 1$, is then normalized by the total spectral energy, that is, $n_f = \frac{E_f}{\sum_{f=0}^{L-1} E_f}, f = 0, \ldots, L - 1$. The entropy of the normalized spectral energy n_f is finally computed according to the equation:

$$H = -\sum_{f=0}^{L-1} n_f \cdot \log_2 (n_f). \tag{4.10}$$

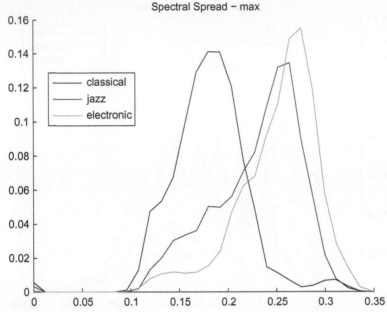

Figure 4.9 Histograms of the maximum value of the sequences of the spectral spread feature, for audio segments from three music genres: classical, jazz, and electronic. The Bayesian error for this three-class classification task, using this single feature, was found to be equal to 41.8%.

The function that implements the spectral entropy is as follows:

```
function En = feature_spectral_entropy(windowFFT, numOfShortBlocks)

% function En = feature_spectral_entropy(windowFFT, numOfShortBlocks)
%
% This function computes the spectral entropy of the given audio frame
%
% ARGUMENTS:
% — windowFFT:       the abs(FFT) of an audio frame
%                    (computed by getDFT() function)
% — numOfShortBins   the number of bins in which the spectrum
%                    is divided
%
% RETURNS:
% — En:              the value of the spectral entropy
%
% number of DFT coefs
fftLength = length(windowFFT);
```

```
% total frame (spectral) energy
Eol = sum(windowFFT.^2);

% length of sub-frame:
subWinLength = floor(fftLength / numOfShortBlocks);
if length(windowFFT)≠subWinLength* numOfShortBlocks
    windowFFT = windowFFT(1:subWinLength* numOfShortBlocks);
end

% define sub-frames:
subWindows = reshape(windowFFT, subWinLength, numOfShortBlocks);

% compute spectral sub-energies:
s = sum(subWindows.^2) / (Eol+eps);

% compute spectral entropy:
En = -sum(s.*log2(s+eps));
```

Figure 4.10 presents the histograms of the standard deviation of sequences of spectral entropy of segments from three classes: music, speech, and others1 (low-level environmental sounds). It can be seen that this statistic has lower

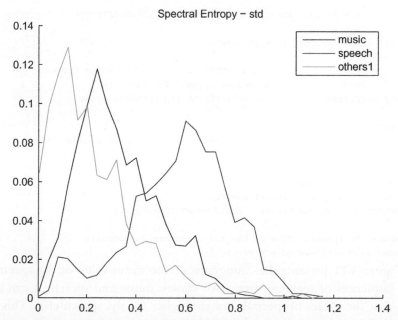

Figure 4.10 Histograms of the standard deviation of sequences of the spectral entropy feature, for audio segments from three classes: music, speech, and others1 (low-level environmental sounds).

values for the environmental sounds, while speech segments yield the highest values among the three classes.

A variant of spectral entropy called chromatic entropy has been used in [23] and [24] in order to efficiently discriminate between speech and music.

4.4.3. Spectral Flux

Spectral flux measures the spectral change between two successive frames and is computed as the squared difference between the normalized magnitudes of the spectra of the two successive short-term windows:

$$Fl_{(i,i-1)} = \sum_{k=1}^{Wf_L} (EN_i(k) - EN_{i-1}(k))^2, \qquad (4.11)$$

where $EN_i(k) = \frac{X_i(k)}{\sum_{l=1}^{Wf_L} X_i(l)}$, i.e. $EN_i(k)$ is the kth normalized DFT coefficient at the ith frame. The spectral flux has been implemented in the following function:

```
function F = feature_spectral_flux(windowFFT, windowFFTPrev)

% function F = feature_spectral_flux(windowFFT, windowFFTPrev)
%
% Computes the spectral flux feature
% ARGUMENTS:
% — windowFFT:              the abs(FFT) of the current audio frame
%                           (computed by getDFT() function)
% — windowFFTPrev:         the abs(FFT) of the previous frame
%
% RETURNS:
% — F:                      the spectral flux value for the input frame
%

% normalize the two spectra:
windowFFT = windowFFT / sum(windowFFT);
windowFFTPrev = windowFFTPrev / sum(windowFFTPrev+eps);

% compute the spectral flux as the sum of square distances:
F = sum((windowFFT — windowFFTPrev).^2);
```

Figure 4.11 presents the histograms of the mean value of the spectral flux sequences of segments from two classes: music and speech. It can be seen that the values of spectral flux are higher for the speech class. This is expected considering that local spectral changes are more frequent in speech signals due to the rapid alternation among phonemes, some of which are quasi-periodic, whereas others are of a noisy nature.

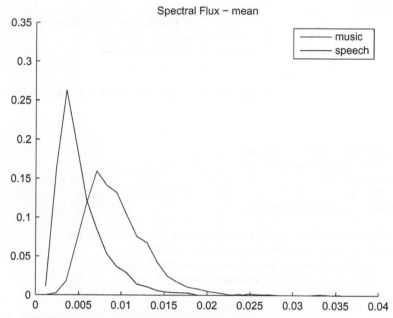

Figure 4.11 Histograms of the mean value of sequences of spectral flux values, for audio segments from two classes: music and speech.

4.4.4. Spectral Rolloff

This feature is defined as the frequency below which a certain percentage (usually around 90%) of the magnitude distribution of the spectrum is concentrated. Therefore, if the mth DFT coefficient corresponds to the spectral rolloff of the ith frame, then it satisfies the following equation:

$$\sum_{k=1}^{m} X_i(k) = C \sum_{k=1}^{Wf_L} X_i(k), \tag{4.12}$$

where C is the adopted percentage (user parameter). The spectral rolloff frequency is usually normalized by dividing it with Wf_L, so that it takes values between 0 and 1. This type of normalization implies that a value of 1 corresponds to the maximum frequency of the signal, i.e. to half the sampling frequency. The MATLAB function that implements this feature, given the DFT spectrum of a frame, is the following:

```
function mC = feature_spectral_rolloff(windowFFT, c)
% function mC = feature_spectral_rolloff(windowFFT, c)
%
% Computes the spectral rolloff feature.
```

```
% ARGUMENTS:
% — windowFFT:            the abs(FFT) of the current audio frame
%                         (computed by getDFT() function)
% — c:                    the spectral rolloff parameter
%
% RETURNS:
% — mC:                   the spectral rolloff value for the input frame
%


% compute total spectral energy:
totalEnergy = sum(windowFFT.^2);
curEnergy = 0.0;
countFFT = 1;
fftLength = length(windowFFT);

% find the spectral rolloff as the frequency position where the
% respective spectral energy is equal to c*totalEnergy
while ((curEnergy≤c*totalEnergy) && (countFFT≤fftLength))
    curEnergy = curEnergy + windowFFT(countFFT).^2;
    countFFT = countFFT + 1;
end
countFFT = countFFT — 1;

% normalization:
mC = ((countFFT—1))/(fftLength);
```

Spectral rolloff can be also treated as a spectral shape descriptor of an audio signal and can be used for discriminating between voiced and unvoiced sounds [7,5]. It can also be used to discriminate between different types of music tracks. Figure 4.12 presents an example of a spectral rolloff sequence. In order to generate this figure, we created a synthetic music track (20 s long) consisting of four different music excerpts (5 s each). The first part of the synthetic track stems from classical music, the second and third parts originate from two different electronic tracks, while the final part comes from a jazz track. It can be easily observed that the electronic music tracks correspond to higher values of the spectral rolloff sequence. In addition, the variation is more intense for this type of music. This is to be expected, if we consider the shape of the respective spectrograms: in the classical and jazz parts, most of the spectral energy is concentrated in lower frequencies and only some harmonics can be seen in the middle and higher frequency regions. On the other hand, in this example, electronic music yields a wider spectrum and as a consequence, the respective spectral rolloff values are higher.

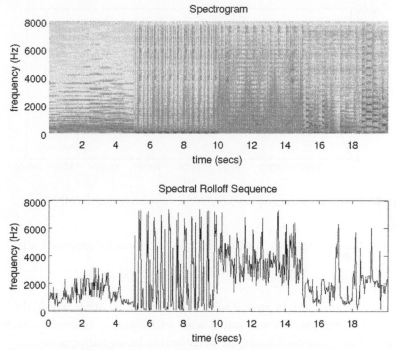

Figure 4.12 Example of the spectral rolloff sequence of an audio signal that consists of four music excerpts. The first 5 s stem from a classical music track. Seconds 5–10 and 10–15 contain two different segments of electronic music, while the last 5 s contain a jazz part. It is readily observed that the segments of electronic music yield higher values of the spectral rolloff feature because the respective spectra exhibit a wider distribution.

4.4.5. MFCCs

Mel-Frequency Cepstrum Coefficients (MFCCs) have been very popular in the field of speech processing [5]. MFCCs are actually a type of cepstral representation of the signal, where the frequency bands are distributed according to the mel-scale, instead of the linearly spaced approach. In order to extract MFCCs from a frame, the following steps are executed:

1. The DFT is computed.
2. The resulting spectrum is given as input to a mel-scale filter bank that consists of L filters. The filters usually have overlapping triangular frequency responses. The mel-scale introduces a frequency warping effect in an attempt to conform with certain psychoacoustic observations [25] which have indicated that the human auditory system can distinguish neighboring frequencies more easily in the low-frequency region. Over the years

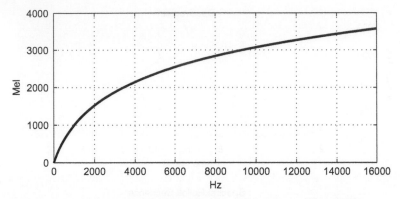

Figure 4.13 Frequency warping function for the computation of the MFCCs.

a number of frequency warping functions have been proposed, e.g.,

$$f_w = 1127.01048 * \log(f/700 + 1)$$

[26]. This equation is presented in Figure 4.13. In other words, the mel scale is a perceptually motivated scale of frequency intervals, which, if judged by a human listener, are perceived to be equally spaced.

3. If $\widetilde{O}_k$, $k = 1, \ldots, L$, is the power at the output of the kth filter, then the resulting MFCCs are given by the equation

$$c_m = \sum_{k=1}^{L} (\log \widetilde{O}_k) \cos\left[m\left(k - \frac{1}{2}\right)\frac{\pi}{L}\right], \quad m = 1, \ldots, L. \quad (4.13)$$

Therefore, according to Eq. (4.13), MFCCs are the discrete cosine transform coefficients of the mel-scaled log-power spectrum.

MFCCs have been widely used in speech recognition [27], musical genre classification [14], speaker clustering [28], and many other audio analysis applications.

In this book we have adopted the implementation proposed by Slaney in [29]. In order to improve computational complexity, we have added a function that precomputes the basic quantities involved in the computation of the MFCCs, i.e. the weights of the triangular filters and the DCT matrix. The function that implements this preprocessing step is the following:

```
function mfccParams = feature_mfccs_init(windowLength, fs)

% function mfccParams = feature_mfccs_init(windowLength, fs)
%
```

```
% This function is used to initalize mfcc quantities
% used in the MFCC calculation
%
% ARGUMENTS:
% — windowLength: the length of the window analysis (in number of samples)
% — fs:          : the sampling frequency (in Hz)
%
% RETURNS:
% — mfccParams   : returns a structure with the mfcc params:
%


% number of cepstral coefficients:
mfccParams.cepstralCoefficients = 13;


% fft resolution:
mfccParams.fftSize = round(windowLength / 2);
% filter parameters:
mfccParams.lowestFrequency = 133.3333;
mfccParams.linearFilters = 13;
mfccParams.linearSpacing = 66.66666666;
mfccParams.logFilters = 27;
mfccParams.logSpacing = 1.0711703;
mfccParams.totalFilters = mfccParams.linearFilters + ...
    mfccParams.logFilters;
mfccParams.freqs = mfccParams.lowestFrequency + ...
    (0:mfccParams.linearFilters−1)*mfccParams.linearSpacing;
mfccParams.freqs(mfccParams.linearFilters+1:mfccParams.totalFilters+2) = ...
    mfccParams.freqs(mfccParams.linearFilters) * ...
    mfccParams.logSpacing.^(1:mfccParams.logFilters+2);
mfccParams.lower = mfccParams.freqs(1:mfccParams.totalFilters);
mfccParams.center = mfccParams.freqs(2:mfccParams.totalFilters+1);
mfccParams.upper = mfccParams.freqs(3:mfccParams.totalFilters+2);
mfccParams.mfccFilterWeights = zeros(mfccParams.totalFilters,mfccParams.fftSize);
mfccParams.triangleHeight = 2./(mfccParams.upper−mfccParams.lower);
mfccParams.fftFreqs = (0:mfccParams.fftSize−1)/mfccParams.fftSize*fs;


for chan=1:mfccParams.totalFilters % for each filter:
    % compute the respective filter weights:
    mfccParams.mfccFilterWeights(chan,:) = (mfccParams.fftFreqs > ...
        mfccParams.lower(chan) & mfccParams.fftFreqs ≤ mfccParams.center(chan)).* ...
        mfccParams.triangleHeight(chan).*...
        (mfccParams.fftFreqs−mfccParams.lower(chan))/...
        (mfccParams.center(chan)−mfccParams.lower(chan)) + ...
        (mfccParams.fftFreqs > mfccParams.center(chan) & ...
        mfccParams.fftFreqs < mfccParams.upper(chan)).* ...
          mfccParams.triangleHeight(chan).*...
          (mfccParams.upper(chan)−mfccParams.fftFreqs)/...
          (mfccParams.upper(chan)−mfccParams.center(chan));
end


% matrix used in the DCT calculation:
mfccParams.mfccDCTMatrix = 1/sqrt(mfccParams.totalFilters/2)*...
```

```
    cos((0:(mfccParams.cepstralCoefficients-1))' * ...
     (2*(0:(mfccParams.totalFilters-1))+1) * pi/2/mfccParams.totalFilters);
mfccParams.mfccDCTMatrix(1,:) = mfccParams.mfccDCTMatrix(1,:) * sqrt(2)/2;
```

The above function returns a structure that contains the MFCC parameters which are then used to compute the MFCCs. The final MFCC values for each short-term frame (assuming that the DFT has already been extracted) are achieved by calling the following function:

```
function ceps = feature_mfccs(windowFFT, mfccParams)

% This function computes the mfccs using the provided DFT.
% The parameters (DCT, filter banks, etc) need to have been
% computed using the feature_mfccs_init function.
%

earMag = log10(mfccParams.mfccFilterWeights * windowFFT+eps);
ceps = mfccParams.mfccDCTMatrix * earMag;
```

The reader can refer to the code of the `stFeatureExtraction()` function (Section 4.1) to understand how the `mfcc` function must be called in the context of the short-term analysis process. We review this process in the following lines:

```
...
mfccParams = feature_mfccs_init(windowLength, fs);
for i=1:numOfFrames % for each frame
    ...
    frameFFT = getDFT(frame, fs);
    MFCCs = feature_mfccs(frameFFT, mfccParams);
    ...
end
...
```

The MFCCs have proved to be powerful features in several audio analysis applications. For instance, in the binary classification task of speech vs music, they exhibit significant discriminative capability. Figure 4.14 presents the histograms of the standard deviation of the 2nd MFCC for the binary classification task of speech vs music. It can be seen that the discriminative power of this feature is quite high, and that the estimated Bayesian error is equal to 11.8% using this single feature statistic. It is worth noting that, depending on the task at hand, different subsets of the MFCCs have been used over the years. For example, it has become customary in many music processing applications to select the first 13 MFCCs because they are considered to carry enough discriminative information in the context of various classification tasks.

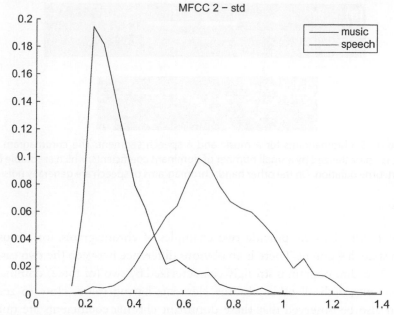

Figure 4.14 Histograms of the standard deviation of the 2nd MFCC for the classes of music and speech. The estimated Bayesian error is equal to 11.8% using this single feature statistic.

4.4.6. Chroma Vector

The chroma vector is a 12-element representation of the spectral energy of proposed in [30]. This is a widely used descriptor, mostly in music–related applications [31,32,33,24].

The chroma vector is computed by grouping the DFT coefficients of a short-term window into 12 bins. Each bin represents one of the 12 equal-tempered pitch classes of Western-type music (semitone spacing). Each bin produces the mean of log-magnitudes of the respective DFT coefficients:

$$v_k = \sum_{n \in S_k} \frac{X_i(n)}{N_k}, \quad k \in 0, \dots, 11, \tag{4.14}$$

where S_k is a subset of the frequencies that correspond to the DFT coefficients and N_k is the cardinality of S_k. In the context of a short-term feature extraction procedure, the chroma vector, v_k, is usually computed on a short-term frame basis. This results in a matrix, V, with elements $V_{k,i}$, where indices k and i represent pitch class and frame number, respectively. V is actually a matrix representation of the sequence of chroma vectors and is also known as the *chromagram* (in an analogy to the spectrogram).

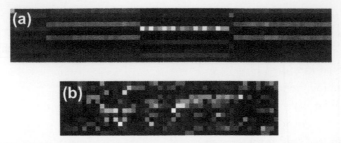

Figure 4.15 Chromagrams for a music and a speech segment. The chromagram of music is characterized by a small number of dominant coefficients, which are stable for a short-time duration. On the other hand, chromagrams of speech are generally noisier.

In Figure 4.15 we provide two examples of chromagrams, for a music and a speech segment. There is an obvious difference between the two cases: the music chromogram is strongly characterized by two (or three) dominant coefficients, with all other chroma elements having values close to zero. It can also be observed that these dominant chroma coefficients are quite 'stable' for a short period of time in music signals. On the other hand, the chroma coefficients are noisier for speech segments (e.g. Figure 4.15b). This difference in the behavior of the chroma vector among speech and music signals has been used in speech-music discrimination applications, e.g. [24].

In our toolbox, the chroma vector of an audio frame is extracted using function `feature_chroma_vector()`.

4.5. PERIODICITY ESTIMATION AND HARMONIC RATIO

Audio signals can be broadly categorized as quasi-periodic and aperiodic (noise-like). The term 'quasi-periodic' refers to the fact that although some signals exhibit periodic behavior, it is extremely hard to find two signal periods that are exactly the same (by inspecting the signal on a sample basis). Quasi-periodic signals include voiced phonemes and the majority of music signals. On the other hand, noise-like signals include unvoiced phonemes, background/environmental noise, applause, gunshots, etc. It has to be noted that there are, of course, gray areas in this categorization, as is, for example, the case with fricative phonemes and some percussive sounds. For the sake of simplicity, we will refer to the quasi-periodic sounds as periodic. For this type of signal, the frequency equivalent of the length of the (fundamental) period of the signal is the so-called *fundamental frequency* and the algorithms that attempt to estimate this frequency are called *fundamental frequency tracking* algorithms. Note that, in the literature, the term 'pitch' is often used interchangeably with the term fundamental frequency. Strictly speaking, pitch represents perceived frequency, like loudness represents perceived signal intensity. However, in many cases, fundamental frequency and pitch coincide, hence the simplification.

A popular, simple technique for estimating the fundamental period is based on the autocorrelation function [8]. According to this method, the signal is shifted and for each signal shift (lag) the correlation (resemblance) of the shifted signal with the original one is computed. In the end, we choose the fundamental period to be the lag, for which the signal best resembles itself, i.e. where the autocorrelation is maximized. We now provide a more detailed description of the algorithmic steps that lead to the estimation of the fundamental frequency based on the autocorrelation function. Our approach follows the guidelines set in the MPEG-7 audio description scheme [34]:

1. Compute the autocorrelation function for frame i, as in the following equation:

$$R_i(m) = \sum_{n=1}^{W_L} x_i(n)x_i(n-m),\qquad(4.15)$$

where W_L is the number of samples per frame. In other words $R_i(m)$ corresponds to the correlation of the ith frame with itself at time–lag m.

2. Normalize the autocorrelation function:

$$\Gamma_i(m) = \frac{R_i(m)}{\sqrt{\sum_{n=1}^{W_L} x_i(n)^2 \sum_{n=1}^{W_L} x_i(n-m)^2}}.\qquad(4.16)$$

3. Compute the maximum value of Γ_i (also known as the *harmonic ratio*):

$$HR_i = \max_{T_{min} \leq m \leq T_{max}} \{\Gamma_i(m)\}.\qquad(4.17)$$

T_{min} and T_{max} stand for the minimum and maximum allowable values of the fundamental period. T_{max} is often defined by the user, whereas T_{min} usually corresponds to the lag for which the first zero crossing of Γ_i occurs. The harmonic ratio can be also used as a feature to discriminate between voiced and unvoiced sounds.

4. Finally, the *fundamental period* is selected as the position where the maximum value of Γ_i occurs:

$$T_0^i = arg\,max_{T_{min} \leq m \leq T_{max}}\{\Gamma_i(m)\}.\qquad(4.18)$$

The *fundamental frequency* is then:

$$f_0^i = \frac{1}{T_0^i}. \tag{4.19}$$

As an example, consider the following periodic signal:

$$x(n) = 0.95 \cos\left(2 \cdot 500\pi\, n\right) + 0.5 \cos\left(2 \cdot 1000\pi\, n\right) + 0.8 \cos\left(2 \cdot 2500\pi\, n\right)$$
$$+ 0.2 \cos\left(2 \cdot 3500\pi\, n\right). \tag{4.20}$$

The number of samples is $W_L = 800$ and the sampling frequency is 16 kHz. Figure 4.16 presents the autocorrelation and normalized autocorrelation functions of this signal. The position of the maximum normalized autocorrelation value defines the fundamental period.

The code that implements the fundamental frequency estimation (along with the harmonic ratio) is the following:

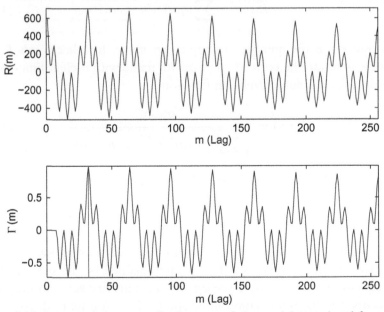

Figure 4.16 Autocorrelation, normalized autocorrelation, and detected peak for a periodic signal. The position (32) of the maximum value of the normalized autocorrelation function corresponds to the fundamental period $\left(\frac{32}{16,000} = 0.002\ \text{s}\right)$. Therefore, the fundamental frequency is $f_0 = \frac{1}{T_0} = \frac{1}{0.002} = 500$ Hz. The maximum value itself is the harmonic ratio.

```
function [HR, f0] = feature_harmonic(window, Fs, M, m0)


%
% function [HR, f0, Gamma] = feature_harmonic(window, Fs, M, m0)
% This function computes the harmonic ratio and fundamental frequency of a
% window
%
% ARGUMENTS
% - window: the samples of the window
% - Fs:     the sampling frequency
% - M:      the maximum T0 (optional)
% - m0:     the minimum T0 (optional)
%
% RETURNS:
% - HR:     harmonic ratio
% - f0:     fundamental frequency
%

if nargin<3
    M=round(0.016*Fs);
end

% compute autocorrelation:

R=xcorr(window);
g=R(length(window));

R=R(length(window)+1:end);
i=2;
if nargin<4
    % estimate m0 (as the first zero crossing of R)
    m0=length(R)+1;
    while i≤length(R)
        if R(i)<0 & R(i−1)≥0
            m0=i;
            break;
        end
        i=i+1;
    end
end

        if M>length(R) M = length(R); end
        % compute normalized autocorrelation:
        Gamma = zeros(M, 1);
        CSum = cumsum(window.^2);
    Gamma(m0:M) = R(m0:M) ./ (sqrt((g*CSum(end−m0:−1:end−M)))+eps);
    Z = feature_zcr(Gamma);
    if Z > 0.15
        HR = 0;
        f0 = 0;
    else
```

```
% compute T0 and harmonic ratio:
if isempty(Gamma)
    HR=1;
    blag=0;
    Gamma=zeros(M,1);
else
    [HR, blag] = max(Gamma);
end
% get fundamental frequency:
f0 = Fs / blag;
if f0>5000 f0 = 0; end
if HR<0.1 f0 = 0; end;
end
```

In Figure 4.17, we demonstrate the discriminative power of the harmonic ratio. In particular, we present the histograms of the maximum value of the harmonic ratio for two classes: speech and others1 (noisy environmental sounds). It can be seen that the adopted statistic takes much higher values for the speech class compared to the noisy class, as is expected. In particular, a binary classifier can achieve up to 88% performance using this single feature.

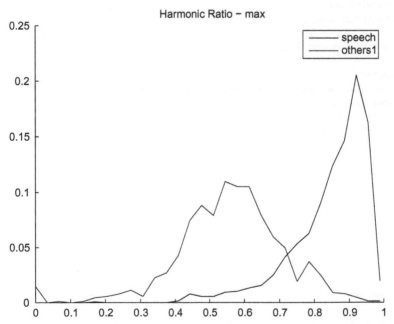

Figure 4.17 Histograms of the maximum value of sequences of values of the harmonic ratio for two classes of sounds (speech and others1). The estimated Bayesian error for this binary classification task is 12.1%.

Remarks

- In `feature_harmonic()` we have used the zero-crossing rate feature in order to estimate the number of peaks in the autocorrelation sequence. Too many peaks indicate that the signal is not periodic. Two other constraints during the computation of the fundamental frequency are associated with the minimum allowable value for the harmonic ratio and the range of allowable values for the fundamental frequency. These values are usually empirically set depending on the application. For example, if we are interested in computing the pitch value of speech signals (e.g. of male speakers), then the range of values of the fundamental frequency should be quite narrow (e.g. 80–200 Hz).
- Fundamental frequency estimation is a difficult task. Therefore, it is no surprise that several spectral methods have also been proposed in the literature, in addition to the popular autocorrelation approach. Some of the exercises at the end of this chapter introduce the key algorithmic concepts of selected frequency-based methods and challenge the reader to proceed with their implementation.

4.6. EXERCISES

1. (D03) Write a MATLAB function that:
   (a) Uses `audiorecorder` to record a number of audio segments, 0.5s each. Section 2.6.2 demonstrates how this can be achieved.
   (b) For each recorded segment, computes and plots (on two separate subfigures) the sequences of the short-time energy and zero-crossing rate features. Adopt a 20 ms frame.

   Note that this process must fit in an online mode of operation.

2. (D04) One of the earliest methods in speech processing for the estimation of the fundamental frequency of speech signals has been the work of Schroeder [35]. Schroeder's method was based on the idea that although the fundamental frequency may not correspond to the highest peak in the spectrum of the signal, it can emerge as the common divisor of its harmonic frequencies. As a matter of fact, the method can even trace the fundamental frequency when it is not present in the signal at all. This line of thinking led to the generation from the speech signal of the so-called "frequency histogram" and its generalization, the "harmonic product spectrum." In this exercise, we provide a simplified version of Schroeder's algorithm based on the DFT of the signal. Assume that the

signal is short-term processed and focus on a single frame. Follow the steps below:

- **Step 1**: Compute the log-magnitude of the DFT coefficients of the frame.
- **Step 2**: Detect all spectral peaks and mark the corresponding frequencies. Let $F = \{(f_k, A_k); k = 1, \ldots, L\}$ be the resulting set, where f_k is a frequency, A_k the magnitude of the respective spectral peak, and L the number of detected peaks. To simplify the pairs in F are sorted in ascending order with respect to frequency. *Hint*: You may find it useful to employ MATLAB's function findpeaks() http://www.mathworks.com/help/signal/ref/findpeaks.html) to detect the spectral peaks.
- **Step 3**: Select f_{min} and f_{max} such that $[f_{min}, f_{max}]$ is the allowable range of fundamental frequencies for the task at hand.
- **Step 4**: Initialize a histogram by placing each f_k at the center of a bin, if $f_k \in [f_{min}, f_{max}]$. The height of each bin is set equal to the log-magnitude of the DFT coefficient that corresponds to f_k.
- **Step 5**: For each f_k in F compute its submultiples in the range $[f_{min}, f_{max}]$. Place each submultiple in the correct bin and increase the height of the bin by the log-magnitude of f_k, i.e. of the frequency from which the submultiple originated. The correct bin is chosen based on the proximity of the submultiple to the bin centers.
- **Step 6**: Select the bin with the highest accumulated magnitude. The frequency at the center of the bin is the detected fundamental frequency.

Provide a MATLAB implementation of the above algorithmic description and compare the results with the autocorrelation method that was presented earlier in this chapter. Run your program on a signal from a single speaker, then for a monophonic music signal, and finally on an excerpt from a polyphonic track. In which case is the resulting pitch-tracking sequence the noisiest? Comment on your findings.

3. (D04) In this exercise you are asked to implement a variant of an MPEG-7 audio descriptor, the Audio Spectrum Envelope (ASE), which forms the basis for computing several other spectral descriptors in MPEG-7. The basic idea behind the ASE feature is that the frequency of 1000 Hz is considered to be the center of hearing and that the spectrum is divided into logarithmically distributed frequency bands around 1000 Hz. To compute the proposed variant of the ASE on a short-term frame basis, implement the following steps:

- **Step 1**: At the first step, the resolution, R, of the frequency bands is defined (as a fraction of the octave) using the equation

$$R = 2^k \text{ octaves}, \quad k = -4, -3, \ldots, 0, 1, \ldots, 3.$$

This equation implies that R can take the following eight discrete values: $\frac{1}{16}, \frac{1}{8}, \frac{1}{4}, \frac{1}{2}, 1, 2, 4, 8$. For example, the value $R = \frac{1}{8}$ means that the width of each frequency band is equal to $\frac{1}{8}$th of the octave.

- **Step 2**: The next step is to define the edges of the frequency bands. To this end, it is first assumed that the frequencies 62.5 Hz and 16 kHz are the lower and upper limits of human hearing, respectively. Then, starting from the lower limit of hearing, the left and right edges of each frequency band are computed as follows:

$$lF_m = 62.5 \cdot 2^{(m-1)R},$$
$$rF_m = 62.5 \cdot 2^{mR},$$

where lF_m, rF_m are the left and right edges of the mth frequency band, $m = 1, \ldots, M$; and $M = \frac{8}{R}$ is the number of frequency bands covering the frequency range 62.5 Hz to 16 kHz. For example, if $R = \frac{1}{4}$, then $M = \frac{8}{0.25} = 32$ and the first frequency band covers the frequency range 62.5 Hz to 74.3254 Hz. Similarly, the last frequency band ($m = 32$) covers the frequency range 13, 454 Hz to 16 kHz. Note, that the smaller the value of m, the more narrow the frequency band. However, if you place the band edges on a logarithmic scale, you will readily observe that each band is always $R = \frac{1}{4}$ octaves wide. Furthermore, for this specific example, if $m = 16$ and $m = 17$, the respective frequency bands are 840.8964 Hz to 1000 Hz and 1000 Hz to 11, 892 Hz, i.e. the center of hearing is the right edge of the 16th band and the left edge of the 17th band.

- **Step 3**: Compute the magnitude of the DFT coefficients of the frame. It is, of course, assumed that you have already decided on the frame length. Note that, until the previous step, all computations conformed with the MPEG-7 standard. However, this step has introduced a variation to the implementation of this MPEG-7 descriptor because we are using the magnitude of the DFT coefficients instead of the 'power spectrum' that MPEG-7 offers [34].

- **Step 4**: Assign each DFT coefficient to the respective frequency band. If a coefficient should fall exactly on the boundary between successive bands, then it is assigned to the leftmost band. Associate a total magnitude with each band. Whenever a DFT coefficient is assigned to a band, increase the cumulative magnitude by the magnitude of the DFT coefficient.

- **Step 5**: Finally, repeat the previous step for all coefficients in the range 0 Hz to 62.5 Hz and 16 kHz and $\frac{F_s}{2}$, where F_s is the sampling rate.

The outcome of the above procedure is a vector of $M + 2$ coefficients. This vector can also be treated as a bin representation of the spectrum or as an envelope (hence, the name, ASE), in the sense that it provides a simplified spectrum representation. Obviously, the smaller the value of R the more detailed the resulting envelop, i.e. parameter R provides the means to control the resolution of the representation. After you implement the proposed variant as an m-file, apply it on various sounds of your choice. Experiment with R in order to get a better understanding of the concept of varying spectral resolution. Can you modify function stFeatureExtraction() to add the functionality of this feature?

4. (D03) Another old-but-popular fundamental frequency tracking technique is the Average Magnitude Difference Function (AMDF) [36]. This method was born in the mid-1970s, out of the need to reduce the computational burden of the autocorrelation approach, which is heavily based on multiplications (see Eq. (4.15)). The key idea behind the AMDF method is that if $x(n)$, $n = 0, \ldots, N - 1$, is an audio frame and T is the length of its period in samples, then the sum of absolute differences between the samples of x and the samples of its shifted version by T should be approximately equal to zero (remember that we are dealing with pseudo-periodic signals, hence, the word 'approximately'). We can therefore define the AMDF at time-lag k as

$$\text{AMDF}(k) = \frac{1}{N - k} \sum_{n=0}^{N-1} | x(n) - x(n - k) | . \qquad (4.21)$$

The above equation is computed for all the values of k whose equivalent frequency lies in the desired frequency range. We then seek the global minimum (usually beyond the first zero crossing of the AMDF). The lag which corresponds to this global minimum is the length of the period in samples.

In this exercise you are asked to implement the AMDF function, apply it on signals of your choice, and compare the resulting pitch sequences with those of the autocorrelation-based approach in Section 4.5.

5. (D04) This exercise revolves around a variant of spectral entropy (Section 4.4.2), the chromatic entropy, which was introduced in [24]. To compute this feature on a short-term frame basis, do the following:

- **Step 1**: Let $[0, f_{max}]$ be the frequency range of interest, where f_{max} is the maximum frequency and $f_{max} < \frac{F_s}{2}$, where F_s is the sampling frequency.

- **Step 2**: All computations are then carried out on a mel-scale. Specifically, the frequency axis is warped according to the equation

$$f = 1127.01048 \cdot \log_e \left(\frac{f_l}{700} + 1 \right), \qquad (4.22)$$

where f_l is the frequency on a linear axis.

- **Step 3**: Split the mel-scaled frequency range into bands. The centers, $f_c, c = 0, \ldots, M - 1$, of the bands coincide with the semitones of the equal-tempered chromatic scale. If, for the moment, frequency warping is ignored, the centers follow the equation:

$$f_c = f_0 \cdot 2^{c/12}, \qquad c = 0, \ldots, M - 1, \qquad (4.23)$$

where f_0 is a starting frequency (13.75 Hz in this case) and M is the number of bands. Given that we use a frequency warped axis, we need to combine Eqs. (4.22) and (4.23) to form

$$f_c = 1127.01048 \cdot \log_e \left(\frac{f_0 \cdot 2^{c/12}}{700} + 1 \right), \qquad c = 0, \ldots, M - 1.$$
$$(4.24)$$

- **Step 4**: Compute the DFT of the frame. Assign each DFT coefficient to a band after detecting the closest band center. Remember that the kth DFT coefficient corresponds to frequency $k\frac{F_s}{N}$, where N is the length of the frame. Therefore, on a mel-scale, the kth coefficient is located at

$$1127.01048 \cdot \log_e \left(\frac{k\frac{F_s}{N}}{700} + 1 \right) \text{ mel units.}$$

If a DFT coefficient is assigned to the mth frequency band, its magnitude is added to the respective sum of magnitudes, X_m, of the band.

- **Step 5**: Each X_m is then normalized as follows:

$$n_m = \frac{X_m}{\sum_{m=0}^{M-1} X_m}, \qquad m = 0, \ldots, M - 1, \qquad (4.25)$$

where n_m is the normalized version of X_m.

- **Step 6**: Due to the way Eq. (4.25) is defined, the n_ms can also be interpreted as probabilities. Therefore, in this final step, the entropy, H, of sequence n_m, $m = 0, \ldots, M - 1$, is computed:

$$H = -\sum_{m=0}^{M-1} n_m \log_2 (n_m).$$

Implement the above feature extraction procedure as an m–file and apply it to an excerpt of classical music and to the speech signal of a single speaker. What can you observe in terms of the standard deviation of this feature over the short-term frames of the two segments?

6. (D05) As the reader may have already suspected, a significant body of pitch-tracking literature revolves around the autocorrelation method. An interesting idea that can also be applied in the context of multipitch analysis of audio signals is introduced in [37], where a generalized auto-correlation method is proposed.

In this exercise, you are asked to provide an approximate implementation of the key processing stages of the work in [37]. More specifically, you will first implement the Summary Autocorrelation Function (SACF) of an audio frame and then the Enhanced Summary Autocorrelation Function (ESAFC) of the frame. To this end, let $x(n)$, $n = 0, \ldots, N - 1$ be an audio frame. Use the following steps:

- **Step 1**: Create a lowpass filter L_p, with a cutoff frequency at 1 kHZ.
- **Step 2**: Create a highpass filter, H_p, with a cutoff frequency at 1 kHZ. Both filters should exhibit a 12 dB/octave attenuation at the stop-band.
- **Step 3**: Filter x with L_p. Let x_{low} be the resulting signal.
- **Step 4**: Filter x with H_p. The output of H_p is first half-wave rectified and then lowpass filtered with L_p. Let H_{high} be the resulting signal.
- **Step 5**: Let $| X_{low} |$ and $| X_{high} |$ be the magnitude of the DFT of x_{low} and x_{high}, respectively. Implement the equation:

$$x_{\text{SACF}} = idft(\, | \, X_{low} \, |^k \,) + idft(\, | \, X_{high} \, |^k \,)$$
$$= idft(\, | \, X_{low} \, |^k + | \, X_{high} \, |^k \,), \qquad (4.26)$$

where k is a user parameter and $k <= 2$. The resulting signal, x_{SACF}, is the SACF with k and compare the SACF with the standard auto-correlation. What do you observe when $k = 2$, and what is the effect of $k < 2$?

- **Step 6**: Clip the SACF to positive values, time-scale it by a factor of 2, subtract the result from the original SACF, and clip the result of the subtraction to positive values. What is the effect of this operation?
- **Step 7**: You can repeat the previous step for a time-scaling factor of three, four, etc.

The outcome of **Step 6** (possibly combined with [optional] **Step 7**) is the Enhanced Summary Autocorrelation Function (ESACF). After you implement the ESACF, apply it on a short-term processing basis on a music signal of two instruments and a speech signal of two voices. What are your conclusions if you place the ESACF of each frame in a column of a matrix and display the resulting matrix as an image?

PART Two

Audio Content Characterization

Audio Classification

Contents

Chapter 4 presented well-known audio features along with related mid-term statistics that can be extracted from the audio signal in the context of audio analysis tasks. Our study in Chapter 4 mainly revolved around the histograms of those features as a means to providing hints for the reader as to the discriminative capability of the respective features.

In this chapter, we go one step further. Our goal is to examine the extracted features in conjunction with some of the most widely used classifiers. Obviously, classification algorithms constitute a huge topic that would take a series of books to cover. In this particular chapter, we have chosen to focus on a specific classification task: the classification of audio segments to a set of predefined audio classes (audio segment classification). In other words, we assume that the audio stream has been preprocessed with an audio segmentation algorithm, which has divided the audio signal into a sequence of audio segments. Chapter 6 will present several such segmentation algorithms. For the moment, we assume that the contents of each segment are homogeneous, i.e. only one audio type is expected to be encountered in each audio segment. Consider, for example, the recording of a radio broadcast of classical music. The output of the segmentation stage may yield a series of segments, where each segment contains classical music, speech, or silence. This is like defining a 3-class task and asking the classifier to take a separate decision on a segment-by-segment basis. By the end of the classification stage, the sequence of classification decisions can be post-processed in order to cancel out isolated errors or merge successive segments that have yielded identical class labels.

It is natural for the reader to wonder if the use of homogeneous content is a restriction, and if it makes sense to study audio classification from such a perspective. First of all, even if a segmentation algorithm is not available, it is always possible to segment the audio stream by means of a fixed-length moving window technique. This naive segmentation scheme will inevitably yield some mixed segments to which the classifier will assign incorrect labels with increased error probability. However, some of these errors can be removed at the post-processing stage resulting in an overall acceptable audio content characterization performance. Secondly, the use of homogeneous content allows us to design useful classifiers that exploit the features (low-level and mid-term, presented in Section 4, in ways that are not restricted to the use of histograms.

Before we begin, note that this is not the only chapter of this book that deals with classification techniques. Chapter 7, which focuses on *template matching* and *Hidden Markov modeling*, provides further insight into classification methodology. However, a major difference between the techniques in this chapter and those of Chapter 7 is that the former are based on the statistical averages of the feature sequences, whereas, the latter exploit the temporal evolution of the feature sequence. Furthermore, temporal modeling and related issues demand that the reader become familiar with a somewhat different world of concepts that can also be used in some more 'exotic' tasks, like audio alignment and synchronization. It is, therefore, more convenient for pedagogical purposes to deal with such concepts in a separate chapter.

5.1. CLASSIFICATION FUNDAMENTALS

As described in Chapter 4, feature extraction can be viewed as a pyramid of layers. At the lowest layer, short-term feature extraction techniques generate sequences of feature vectors from the audio stream. At the next (higher) level, a number of statistics are computed over the respective short-term feature sequences. The extracted statistics can then be grouped to form a single feature vector. If D feature sequences have been extracted and S statistics have been computed over each sequence, then the resulting feature vector, say x, is $L = D \times S$-dimensional. In addition, let ω_i, $i = 1, \ldots, N_c$ be the labels of N_c predefined audio classes. Our goal is to estimate the class label, y, of an audio segment that is represented by the feature vector x.

5.1.1. The Bayesian Classifier

A probabilistic approach to the classification task is provided by the *Bayesian classifier*, which bases its decision on the posterior probabilities,

$$P(\omega_i|x), i = 1, \ldots, N_c,$$

of the class labels, ω_i, given the feature vector x. If these posterior probabilities are available, the feature vector is assigned to the class that produces the maximum posterior probability, i.e.

$$y = \arg \max_{\omega_i, i=1,\ldots,N_c} P(\omega_i|x).$$

In general, each $P(\omega_i|x)$, $i = 1, \ldots, N_c$ is a probability density function because the elements of x usually take continuous values. The Bayesian classifier is known to be optimal with respect to the probability of the classification error and it therefore provides a performance bound for every classifier.

To get a better understanding of the *Bayesian error*, consider a 2-class problem, e.g. a speech (ω_1) vs music (ω_2) audio segment classification task. If $p(\omega_1|x) > p(\omega_2|x)$, then the feature vector will be assigned to class ω_1 and the probability of error will be equal to the posterior probability of class ω_2, i.e. equal to $p(\omega_2|x)$. Similarly, if $p(\omega_1|x) < p(\omega_2|x)$, the probability of error is equal to $p(\omega_1|x)$. In addition, if $p(x)$ is the probability of the occurrence of x, then we can average over the entire feature space and end up with the following equation for the probability, P_{err}, of the classification error (Bayesian error):

$$P_{err} = \int_{R_1} P(\omega_1|x)p(x)dx + \int_{R_2} P(\omega_2|x)p(x)dx, \tag{5.1}$$

where R_1 stands for all the feature vectors that were erroneously classified to ω_2 (although they belong to ω_1) and, similarly, R_2 stands for all feature vectors that were erroneously classified to ω_1 (although they belong to ω_2).

A practical issue with the Bayesian classifier is that it is not readily applicable in a real-world scenario because the posterior probability density functions are, in general, intractable. A more convenient form of the Bayesian classifier can be derived if we rewrite $P(\omega_i|\boldsymbol{x})$ using the Bayesian rule of probability, as follows:

$$P(\omega_i|\boldsymbol{x}) = \frac{P(\boldsymbol{x}|\omega_i) \cdot p(\omega_i)}{p(\boldsymbol{x})}. \tag{5.2}$$

The denominator of Eq. (5.2) is the same irrespective of the class and it can therefore be ignored when we are looking for the class that maximizes the posterior probability. In other words, we seek the best class label, y, so that

$$y = \arg \max_{\omega_i, i=1,\ldots,N_c} P(\boldsymbol{x}|\omega_i) \cdot p(\omega_i). \tag{5.3}$$

The probabilities $p(\omega_i)$, $i = 1, \ldots, N_c$ are also known as the *a priori* class probabilities (or *class priors*) because they encode our knowledge of the problem if no feature vector is given. In other words, a class prior is the probability that the respective class is encountered in the context of the classification task being studied. In many applications, the priors can be estimated by means of frequencies of events or the classes can be assumed to be equiprobable and consequently, the classification becomes entirely dependent on the conditional probability density functions (PDFs) $P(\boldsymbol{x}|\omega_i)$. These pdfs capture the likelihood of the feature vectors given each class. Although the complete knowledge of the pdfs $P(\boldsymbol{x}|\omega_i)$, $i = 1, \ldots, N_c$ is in general an impossible task, these pdfs are more attractive because they can be approximated using various techniques, e.g. Maximum Likelihood Estimation, k-NN density estimation, and so on. If a satisfactory approximation of these conditional pdfs is derived, then an approximation of the Bayesian classifier can be put in use. This means, that the basic Bayesian classification mechanism can be preserved and this is why, in the literature and throughout this book, we use the term Bayesian error even when only approximations of the conditional pdfs are used. For a more detailed treatment of the theory around the Bayesian classifier, the reader is referred to [5, 38, 39]. A MATLAB approach to the subject is given in [40].

The Bayesian classifier, although optimal with respect to the probability of error, requires precise knowledge of the respective probability density functions, which are generally intractable. However, due to the need for the design of working solutions, it is often useful in practice to assume that the features are statistically independent and that the pdf of each feature (per class) can be parametrized using well-studied density functions, like the Gaussian pdf. Under these assumptions, it is possible to resort to a simple (naive) approximation of the Bayesian classifier, known as the *naive Bayes classifier* [5, 41, 42] (see also the related exercise at the end of this chapter).

5.1.2. Classifier Training and Testing

Before we proceed with the description of various classifiers, which can be readily applied to the analysis of audio segments, we briefly explain an important stage which is of paramount importance in the design of most classifiers, namely the *training* stage. We start with a description of the training stage as we will frequently refer to training issues during our presentation of classification methods. Note that another important stage, the *testing stage* will be thoroughly treated towards the end of this section.

To begin with, it is assumed that the designer of the classifier has access to a set of *training data*, a set of feature vectors whose class labels are already known. The training dataset is usually the product of a time-consuming procedure which, in the case of audio signals, involves data gathering, listening tests, manual annotation (tagging), and organization of the annotated audio data to data structures in order to facilitate data exchange. In general, the quality of the training set determines the quality of the resulting classifier. In other words, high-quality training data are indispensable. Over the years, some datasets have become a point of reference for researchers and engineers alike, because they can be used to provide a comparative study of methods and tools. For example, the TIMIT Acoustic-Phonetic Continuous Speech Corpus [43] has been a standard corpus of reference for speech recognition techniques. As another example, in the last decade, in the field of music information retrieval, a popular dataset used for music genre classification is the *GTZAN Dataset* [15]. Appendix C provides a list of widely used audio datasets for several classification tasks.

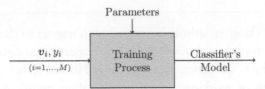

Figure 5.1 Generic diagram of the classifier training stage. The feature vectors, v_i, $i = 1, \ldots, M$, of the training dataset and the respective class labels, y_i, have to be available before the training procedure begins. Furthermore, the designer of the system usually needs to make certain decisions related to key parameters of the classifier. The output of the training stage depends on the type of classifier that has been adopted.

Once the training dataset has been made available, we adopt a classification method and try to tune its parameters so that the classifier 'learns' the training dataset. This is known as the *training stage*, a fundamental concept in *machine learning*. The number of free parameters which are involved in the training stage depends on the complexity of the classification task and the classification algorithm that has been adopted. For example, if a linear classifier is used in a 2-class task where the classes are separable, the goal of the training stage is to determine the hyperplane that separates the training data of the two classes in the feature space. Most classifiers (especially the most sophisticated ones) may demand from the user to make various decisions with respect to the design of the classifier before the actual training procedure begins. Such decisions may refer to a distance metric, order of a polynomial, or choice of a kernel function, and, in any case, require that the designer of the classifier is also familiar with the theory on which the respective classifier is based.

A generic diagram of the training stage is shown in Figure 5.1. One way to measure how well the classifier has learned the training set is to compute the *training error*, which is the percentage of vectors of the training dataset that remain erroneously classified at the end of the training stage. Ideally, we would like the training error to be equal to zero, however, there is a subtle issue here: we would also like our classifier to be capable of *generalizing*. This means that we want the classifier to perform well (classify correctly) on data outside the training dataset. Unfortunately, a zero training error does not guarantee generalization and can be associated with a phenomenon known as *overfitting*, which means that the classifier has learned very well the training set but performs poorly on unknown feature vectors. We defer a more detailed treatment of errors until the second half of this chapter.

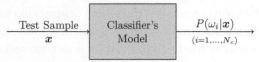

Figure 5.2 Diagram of the classification process. The classifier takes as input a feature vector and its output is the estimated class label and (possibly) the associated posterior probabilities.

After the classifier has been trained, it can be used to classify features that belong to the *testing dataset*. These are features whose class label needs to be determined by the classifier, and the resulting classification decisions must be compared with the true class labels of the data. When such feature vectors are fed to the classifier, we pretend that we are not aware of their true class label; we only use our *ground truth* knowledge when we *evaluate (assess)* the performance of the classifier. It is important to note that the training and test datasets must not overlap, otherwise the evaluation of the performance of the classifier cannot be trusted. Figure 5.2 presents an overview of the classification process of a test sample. The input to the classification model is the feature vector that represents the unknown data (e.g. a feature vector extracted from an audio segment). The output of this process is the estimated class label. Furthermore, depending on the classifier, estimates of the posterior probabilities, $P(\omega_i|\mathbf{x})$, $i = 1, \ldots, N_c$, which are associated with the classifier's decision, may also be returned. The latter can be very useful when we want to measure the confidence of the classifier when it makes its decisions.

5.1.3. Multi-Class Problems

If the number of classes, N_c, is equal to two, the respective classification task is called *binary*. In binary problems, we usually use 0 and 1 for the two class labels although this is not a restrictive choice. For example, $-1, +1$ can be used instead. If more than two classes are involved, we are dealing with a *multi-class* task. Some classification methods are straightforward to use in both a binary and a multi-class scenario as, is the case with the popular k-Nearest-Neighbor (k-NN) classifier [44, 5, 39]. This is because the core mechanism of the classification algorithm can be readily extended to cover the case of more than two classes, both from a training and testing perspective. On the other hand, several classification methods, like the Support Vector Machines (SVMs), which will be described later in the chapter, are designed to operate in a binary classification context and the system designer has to resort to more sophisticated design solutions in order to deal with multi-class situations.

To this end, a widely used design solution is to train several binary classifiers and combine them, so that they provide a multi-class result. In other words, the multi-class problem is treated as a combination of binary classification tasks, a technique also known as *binarization* [45]. Following this line of thinking, two popular approaches are frequently encountered in the literature, namely the *One-vs-All* and *One-vs-One* methods:

- **One-vs-All (OVA)** According to this method (also known as one-vs-rest—OVR), one N_c binary classifier is employed per class. The goal is to train the individual classifiers to discriminate between the samples of the respective class (*positive examples*) and the samples of all the other classes (*negative examples*). Each binary classifier produces a 'soft' output, which can be also interpreted as a confidence score. During the testing phase, the unknown sample is classified to the class which is associated with the binary classifier that has produced the maximum (positive) soft output. Despite its simplicity, this scheme has proved to exhibit significant discriminative capability in various multi-class problems [46].

- **One-vs-One (OVO)** This is another type of classifier binarization, also known as pairwise or round robin classification [45]. It is based on transforming the initial multi-class task into $\frac{N_c(N_c-1)}{2}$ binary classification problems, where each binary problem involves two classes. The total number of classifiers is therefore equal to the number of all possible class pairs. A simple way to combine the individual binary classification decisions to a global decision is via a voting scheme. Specifically, the decision of each pairwise classifier increases by one point the score of the class that won. In the end, the global decision is made based on the class that has accumulated the highest score.

Note: Multi-class classification should *not* be confused with *multi-label* classification [47], according to which, a sample can simultaneously belong to more than one class, as is the case with document (text) classification applications. For instance, it is possible that a research paper lies both in the audio analysis and pattern recognition domains, hence the need to assign multiple labels to its content.

5.2. POPULAR CLASSIFIERS

We start our description of selected classifiers with the famous *k-nearest-neighbor classifier (k-NN classifier)* and proceed with the perceptron and the more sophisticated decision trees and support vector machines (SVMs). Obviously, this is just a very small subset of the classifiers that have been proposed in the literature but serves well our purpose to focus on selected methods which are both popular and representative of the wealth of techniques that are available. We have avoided lengthy theoretical descriptions of the classifiers and have instead made an attempt to highlight the key ideas behind the algorithms being studied. Furthermore, the evaluation techniques and case studies that follow this section aim at examining the presented methods in the broader context of actions that need to be performed when we design useful classification systems. As a final remark, remember that a category of classifiers that revolves around temporal modeling will be further presented in Chapter 7.

5.2.1. The *k*-Nearest-Neighbor Classifier (*k*-NN)

Despite its simplicity, the *k*-nearest-neighbor classifier is well tailored for both binary and multi-class problems. Its outstanding characteristic is that it does not require a training stage in the strict sense. The training samples are rather used directly by the classifier during the classification stage. The key idea behind this classifier is that, if we are given a test pattern (unknown feature vector), x, we first detect its *k*-nearest neighbors in the training set and count how many of those belong to each class. In the end, the feature vector is assigned to the class which has accumulated the highest number of neighbors. Therefore, for the *k*-NN algorithm to operate, the following ingredients are required:

1. A dataset of labeled samples, i.e. a training set of feature vectors and respective class labels.
2. An integer $k \geq 1$.
3. A distance (dissimilarity) measure.

Let us now go through the *k*-NN algorithm in more detail. In the first step, the algorithm computes the distance, $d(x, v_i)$, between x and each feature vector, $v_i, i = 1, \ldots, M$, of the training set, where M is the total number of training samples. The most common choice of distance measure is the

Euclidean distance, which is computed as:

$$d(x, v_i) = \sqrt{\sum_{j=1}^{D} (x(j) - v_i(j))^2}, \qquad (5.4)$$

where D is the dimensionality of the feature space. Another popular choice of distance measure is known as the *Mahalanobis distance* [5]. After $d(x, v_i)$ has been computed for each v_i, the resulting distance values are sorted in ascending order. As a result, the k first values correspond to the k closest neighbors of the unknown feature vector. Now, let k_i be the number of training vectors among the k neighbors of x that belong to the ith class, $i = 1, \ldots, N_c$. The unknown vector is then classified to the class which corresponds to the maximum k_i. Furthermore, the posterior probability $P(\omega_i | x)$ can be estimated as:

$$P(\omega_i | x) = \frac{k_i}{k}, i = 1, \ldots, N_c. \qquad (5.5)$$

Therefore, from a Bayesian perspective, the algorithm assigns the unknown sample to the class that corresponds to the maximum estimated posterior probability.

It is important to note that the k-NN classifier can operate directly in a *multi-class* environment. This is because its algorithmic steps, including the estimation of posterior probabilities at the output of the algorithm, are not restricted by the number of classes that are involved [44]. The multi-class operation of the k-NN classifier can also be considered equivalent to using an ensemble of binary k-NN classifiers (one per class) and combining their classification decisions according to the one-versus-all approach of Section 5.1.3.

The performance of the k-NN classifier has been extensively studied in the literature. An interesting theoretical finding is that if $k \to \infty$, $M \to \infty$, and $\frac{k}{M} \to 0$, then the classification error approaches the Bayesian error [5]. In other words, if both k and M approach infinity and k is infinitely smaller than M, then the k-NN classifier tends to behave like the optimal (Bayesian) classifier with respect to the classification error. In practice, the larger the dataset, the more satisfactory the performance of the k-NN algorithm. Concerning parameter k (the number of neighbors), it can be stated that it is tuned after experimentation with the dataset at hand. In general, small values are preferred by also taking into account the size of the dataset (so that $\frac{k}{M}$ remains as small as possible). Another important issue is the computational complexity of the k-NN classifier which can be prohibitively high when the volume of the dataset is really large (mainly due to the number of Euclidean distances that need to be computed). Over the years, several remedies to this computational issue have been proposed in the literature, e.g. [48, 49].

Notes: The nearest neighbor rule can be also used as an estimator of the conditional pdfs $P(x|\omega_i)$, $i = 1, \ldots, N_c$ and of the probability density function $P(x)$.

To understand how this works, we need to focus on the training samples of a specific class, say ω_i (the procedure is repeated for each class, separately). In order to estimate $P(x|\omega_i)$, we grow a (hyper)sphere around point x, until it encloses k_i training samples of the class. Then, $P(x|\omega_i)$ can be estimated according to the equation:

$$P(x|\omega_i) = \frac{k_i}{M_i V_i}, \qquad (5.6)$$

where M_i is the total number of training samples of class ω_i and V_i is the volume of the (hyper)sphere that encloses the k_i points of the class.

In a similar manner, if we take into account all training samples and drop class labels, we can estimate $P(x)$, i.e. the probability to encounter feature vector x in the feature space of the problem at hand, as follows:

$$P(x) = \frac{k}{MV}, \qquad (5.7)$$

where V is the volume of the (hyper)sphere that encloses the k points of the training set. Obviously, $M = \sum_{i=1}^{M} M_i$, where M is the size of the training set. Furthermore, the prior probabilities of the classes can be estimated as frequencies of occurrence of events, i.e.

$$P(\omega_i) = \frac{M_i}{M}. \qquad (5.8)$$

Finally, we can employ the Bayesian Rule of probabilities to combine Eqs. (5.6–5.8) and derive estimates of the posterior probability density functions, $P(\omega_i|x)$, $i = 1, \ldots, N_c$.

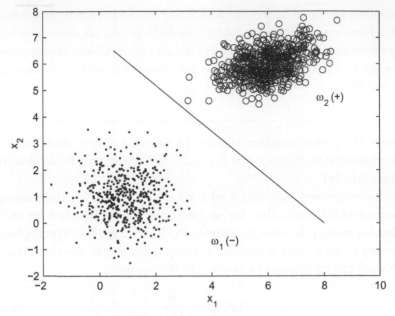

Figure 5.3 Linearly separable classes in a two-dimensional feature space.

5.2.2. The Perceptron Algorithm

Consider a binary classification problem in a two-dimensional feature space, with *linearly separable* classes. Figure 5.3 demonstrates a related example. The term '*linearly separable*' refers to the fact that a line in a two-dimensional feature space, a plane in a three-dimensional scenario, or a hyperplane in the case of more than three dimensions, can separate the two classes without producing any misclassifications at all. It is easier to visualize this scenario on the two-dimensional plane of Figure 5.3, where it can be seen that the selected line leaves the data of each class on a separate half-plane. It can be also observed that certain points of class ω_2 lie very close to the line, but we actually do not care at the moment. Our concern is to simply draw a line that separates the two classes although it can be easily understood that there exists an infinite number of such lines.

Now, let us focus on the equation of the line, i.e.

$$g(x) = w_1 \cdot x_1 + w_2 \cdot x_2 + w_0 = 0, \tag{5.9}$$

where $x = [x_1 \; x_2]^T$ is the two-dimensional feature vector (column vector) and w_1, w_2, w_0 are real numbers which define the line. If we investigate the geometric relationship between each class and the line, we observe that, in the example of Figure 5.3, if any sample of ω_1 is given as input to function $g(x)$, a negative result will be generated. Similarly, the samples of class ω_2 will yield positive values. Therefore, ω_1 lies on the negative half-plane and ω_2 on the positive half-plane with respect to the line. Due to its ability to discriminate between classes, function $g(x)$ is also called a *linear discriminant function*. The reader who is interested in delving into the world of discriminant functions will soon realize that this is just a simple case of discriminant and that a large number of such functions, of increasing complexity, are available in the literature.

It is important to realize that if a sample of ω_1 is misclassified by the line, the discriminant will produce a positive result, i.e. a result which has the sign of the wrong class (positive sign). This result can be interpreted as the cost of the failure. Similarly, a misclassified sample of ω_2 will result in an undesirable negative value, i.e. the result will again have the sign of the wrong class (negative sign this time). In any case, we can multiply the result of each failure with the sign of the wrong class. In this way, the cost will always be positive.

Imagine now, that we need an algorithm that takes as input the training dataset in question and returns a line that is capable of separating the two classes. We are not posing any other restrictions on the solution. For example, we could demand that it does not lie very close to any one of the two classes, but this is not our concern here. A celebrated algorithm that can fulfill our requirement is the *perceptron* algorithm. The basic steps of this algorithm are as follows:

1. Start with a random line (initialization).
2. Determine all misclassified samples based on the current line. Each mis-classified sample is associated with a cost (positive quantity), that was previously explained. The accumulated cost is the sum of the individual costs.
3. Form set Y to include all misclassified samples.
4. If Y is empty, the algorithm terminates. Otherwise, the equation of the line is updated to deal with the accumulated cost that has been generated by the classification failures.
5. Repeat from step 2.

In order to derive a mathematical formulation of the perceptron learning rule, we first group w_1 and w_2 of Eq. (5.9) into a column vector, $w = [w_1 w_2]^T$, and write Eq. (5.9) as:

$$g(x) = x^T \cdot w + w_0 = w^T \cdot x + w_0 = 0, \qquad (5.10)$$

where ^{T} stands for the transpose operator. We then augment x by one dimension and write Eq. (5.11) in pure vector form:

$$g(\hat{x}) = \hat{x}^T \cdot \hat{w} = \hat{w}^T \cdot \hat{x} = 0, \tag{5.11}$$

where $\hat{x} = [x_1 \ x_2 \ 1]^T$ and $\hat{w} = [w_1 \ w_2 \ w_0]$. In other words, we have made the assumption that the classes lie on the plane with a z-coordinate equal to 1. Obviously, this is not a restriction; it is just a trick to rewrite the equation of the line in a more convenient form.

If we drop the hat notation to simplify the presentation, then, based on Eq. (5.11), we can define the cost function of step 2 as

$$J(w) = \sum_{x \in Y} \delta_x w^T x, \tag{5.12}$$

where $\delta_x = +1$ or $\delta_x = -1$, depending on the sign of $w^T x$, so that the respective cost is always a positive quantity. Obviously, if $J(w) > 0$ at least one classification failure has occurred.

The perceptron rule that updates the weights of the discriminant is

$$w(t+1) = w(t) - \rho_t \sum_{x \in Y} \delta_x x, \tag{5.13}$$

where $w(t+1)$ is the weight vector of the updated line (hyperplane in general), $w(t)$ is the current line (hyperplane), and ρ_t is a parameter that controls the speed of convergence of the training algorithm. Usually, ρ_t is chosen to be a constant, e.g. $\rho = 0.7$. In general, the smaller the value of ρ, the slower the convergence of the algorithm. It can be proved that the perceptron learning algorithm converges in a finite number of steps, if the classes are linearly separable. In the case of classes that are not linearly separable, the training algorithm will run infinitely and will never achieve a zero accumulated cost. This means that there will always be classification failures in the training set. The perceptron learning algorithm lies in a broad category of algorithms which converge to a solution (discriminant function) based on *cost function optimization*.

After the training algorithm has converged to a hyperplane that separates the two classes, the respective discriminant function can be used to classify unknown samples based on the sign of the result that it produces, i.e. if $g(x) < 0 (> 0)$ the sample is assigned to $\omega_1 (\omega_2)$. Equivalently, if the output of $g(x)$ is given as input to the sign function, the output of the classification step will be $+1$ or -1. The sign function, $f(u)$, is defined as

$$f(u) = \begin{cases} +1 & \text{if } u \geq 0 \\ -1 & \text{if } u < 0 \end{cases},$$

and is also known as the *activation function*. Therefore, the output of the classifier is the output of the synthesis of two functions, the discriminant function and the activation function. This type of processing is similar to the functionality of the *neurons (perceptrons)* of the human brain, hence the name of the algorithm. The perceptron can be the building block of more complicated structures known as *neural networks*, which can provide classification solutions for complicated tasks (the requirement for linearly separable classes can be raised). For more details on neural networks, the interested reader is referred to any textbook on pattern recognition, machine learning, or related disciplines, e.g. [5, 38, 39].

In older MATLAB versions, the perceptron was implemented in the Neural Network toolbox with function newp(), which is still available but considered to be obsolete. In the latest versions, the perceptron functionality has been embedded in functions that deal with the broader case of neural networks, because the perceptron can be treated as the simplest case of neural network (just one single neuron). At the end of this section, we provide programming exercises whose goal is to help the reader become familiar with the perceptron training algorithm and the perceptron as a classifier.

As a final note, even if two classes are not linearly separable, it is not always a bad idea to use a perceptron as a classifier; sometimes we can tolerate a small amount of error, which, after all, can be due to the existence of outliers in the training set (e.g. noisy samples). In such cases, we need to insert a mechanism into the training algorithm, so that the training procedure terminates, if, after a number of epochs, there still exist classification failures (see respective exercise).

5.2.3. Decision Trees

Decision trees have been widely used in the fields of machine learning and data mining over the years. Two important types of decision trees are the *classification trees* (the predicted outcome is a class label) and the *regression trees* (the predicted outcome is a real number). The term 'Classification and Regression Tree' (*CART*) refers to both types of decision tree [50].

In this chapter, we are mainly interested in classification trees, whose mode of operation is based on making sequential decisions on individual features. [51, 5, 52]. More specifically, in a classification tree:

- Each internal node (non–leaf node) represents the *binary question* 'is the value of feature F larger than T?' This has the effect of *splitting* the feature space based on the examined feature.
- Each leaf is assigned a class label.
- A test sample is classified by navigating the tree from its root node to a leaf (final classification decision), based on the outcome of the 'answers' at the internal nodes.

A simple decision tree is shown in Figure 5.4, for a 3-class (ω_1, ω_2, and ω_3) task with three features (x_1, x_2, and x_3). The $\star$ symbol annotates the decisions that lead to the leaf node. Note that in some cases, certain features may not contribute to the final classification decision; in this figure, for example, x_3 is not visited at all.

A geometric interpretation of the decision trees is that they can be treated as a collection of hyperplanes, where each hyperplane is orthogonal to the respective feature axis. In other words, each internal node adds a decision hyperplane, which splits the respective feature dimension in a binary manner.

Figure 5.5 demonstrates a decision tree that has been employed in a 4–class problem. The features, x_1 and x_2, follow a two–dimensional Gaussian distribution (with different mean vectors and covariance matrices) in each class (A, B, C, or D). Figure 5.5b provides the geometric interpretation of the decision tree: each rectangular area is defined according to the hyperplanes that correspond to the questions at the internal nodes of the tree. In this case, the decision tree is rather simple. At the first level, feature x_1 is used to split the feature space into the class pairs (A, C) and (B, D). The second level uses feature x_2 to discriminate between the classes of each pair. The

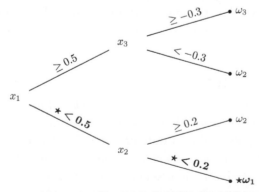

Figure 5.4 Decision tree for a classification task with 3-classes (ω_1, ω_2, ω_3) and three features (x_1, x_2, x_3). The $\star$ symbol annotates the path that leads to the classification decision for the test sample (0, 0, 0). Note that the third feature (x_3) is not visited in this example.

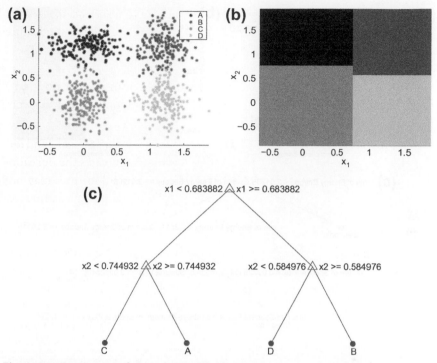

Figure 5.5 Decision tree for a 4-class task with Gaussian feature distributions in the two-dimensional feature space. The operation of the decision tree is equivalent to splitting the feature space based on hyperplanes, which are orthogonal to the respective axes. (a) Distribution of the feature vectors, (b) division of the feature space according to the hyperplanes, and (c) decision tree for the classification task being studied.

example was generated with function demoCART_general, which can be found in the software library that accompanies this book.

Note: CARTS are implemented in the MATLAB Statistics Toolbox. In order to train a decision tree, function classregtree() can be used. Its main input arguments are the feature vectors of the training set and the respective class labels. The training procedure is also capable of handling categorical features and is tolerant to missing values. After the tree has been trained, function eval() can be used to classify unknown samples. Furthermore, function view() helps to build a visualization of the decision tree. Our functions, demoCART_general and demoCART_audio, serve to demonstrate the use of the aforementioned functions of the Statistics Toolbox.

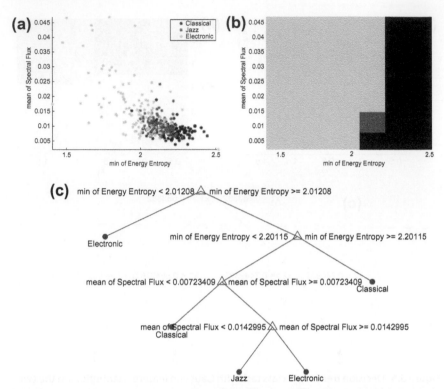

Figure 5.6 Decision tree for a musical genre classification task with two feature statistics (minimum value of the entropy of energy and mean value of the spectral flux). Note that the samples of the Jazz genre are narrowly distributed in the selected feature space: this leads the rules of the decision tree to define a smaller area for this genre. (a) Feature distribution, (b) resulting division of the feature space, and (c) decision tree for the given classification task.

Figure 5.5 presents an example of a trained decision tree classifier on a synthetic set of two-dimensional data. The demoCART_audio script demonstrates an example with real audio features, focusing on two feature statistics of the previously described 3-class musical genre classification task. The two features are the minimum value of the entropy of energy and the mean value of the spectral flux sequence. The respective results are shown in Figure 5.6. Due to the increased complexity of the classification task, the decision tree has more levels compared to the one in Figure 5.5 and performs a more complicated division of the feature space.

Decision trees exhibit certain advantages, some of which are:

• They are self-explanatory and easy to interpret. Their structure can be easily converted to a set of simple rules that can be also interpreted by humans.

- They can easily handle *nominal* features.
- They can handle datasets with missing values.
- No assumptions are needed regarding the feature distributions when training a decision tree classifier. Therefore, decision trees can be considered as a non-parametric approach.

An important drawback of decision trees is their sensitivity to noise, which can make them *unstable*, depending on the training dataset [51]. Furthermore, for the case of complex feature spaces and hard classification tasks, the resulting decision trees can be very complex.

Note: Decision trees can be directly used in a multi-class scenario (Section 5.1.3), as was also the case with the *k*-Nearest-Neighbor classifier. This ability of the decision trees to handle multi-class problems has been demonstrated in both examples of this section (Figures 5.5 and 5.6).

5.2.4. Support Vector Machines

Support Vector Machines (SVMs) are state-of-the-art classifiers that have been successfully employed in numerous machine learning fields [5, 38, 39, 53, 54]. They are based on the key observation that, for the simple case of linearly separable classes, the optimal decision hyperplane (line on the two-dimensional space) is the one that maximizes the 'margin' among the training data of the two classes. To this end, supporting hyperplanes are used, parallel to the decision hyperplane, in order to define the margin that minimizes the classification error. This rationale is extended to cover the non-separable case [53] by permitting misclassified training data.

A description of the theory that governs the operation of SVMs is beyond the scope of this book; we will rather provide a user-centric explanation of the crucial SVM parameters, so that the user can start experimenting with the respective m-files:

- Type of kernel: According to the SVM methodology, a kernel function is used in order to map the feature vectors to the 'kernel space' [53, 54]. Typical kernels are linear, polynomial (of order 3 or higher), and radial basis function (RBF) kernel, etc.
- Kernel properties: depending on the selected kernel, certain kernel parameters need to be set for the training and evaluation of the SVM classifier. For example, in the case of a polynomial kernel, the order of the polynomial must be specified by the user.

- Constraint parameter, C. It is related to the cost function of the SVM training procedure. As the value of C increases, the cost of misclassified samples also increases. Care must be taken so that overfitting is avoided.

The Bioinformatics Toolbox of MATLAB provides a set of functions for training and evaluating SVM classifiers:

- svmtrain: It takes as input (a) a matrix with training vectors (one vector per row, one feature dimension per column), and (b) a vector of respective class labels. It returns a MATLAB structure that represents the trained SVM classifier.
- svmclassify: It receives as input (a) the SVM structure that has been returned by svmtrain, and (b) one or more samples to be classified.

Both functions can plot the results of the training and testing stages, by setting the Showplot flag equal to true. Note that the class labels can be character strings (instead of integers). The following code demonstrates how to use the SVM classifier for a speech vs music classification task:

```
% Load the kNN speech–music model
% (only the features will be used here for training / evaluating the SVM):
load modelSM;

% Get the features to matrix (rows correspond to samples):
F       = [Features{1}, Features{2}]';

% Define the class labels (speech – music):
Labels  = cell(size(Features{1}, 2) + size(Features{2}, 2), 1);
Labels(1:size(Features{1}, 2))       = {'Speech'};
Labels(size(Features{1}, 2)+1:end) = {'Music'};
numOfSamples = length(Labels);                % total number of samples
randPerm     = randperm(numOfSamples);        % rand indices

% training samples and labels (half of the data):
Ftrain      = F(randPerm(1:numOfSamples/2), :);
LabelsTrain = Labels(randPerm(1:numOfSamples/2));

% testing samples and labels (remaining half):
Ftest       = F(randPerm(numOfSamples/2+1:end), :);
LabelsTest  = Labels(randPerm(numOfSamples/2+1:end));

% train the SVM classifier (default values used):
svmStruct = svmtrain(Ftrain, LabelsTrain);

% test the SVM classifier:
LabelsFound = svmclassify(svmStruct, Ftest);

% compute accuracy
% (number of correctly classified samples / total number of testing samples):
Accuracy    = length(find(strcmp(LabelsFound, LabelsTest))==1) ...
    / length(LabelsFound)
```

The code goes through the following steps:

- Loads the dataset of feature vectors.
- Randomly splits the dataset into two equally sized subsets, one for training and one for testing. This validation approach is further described in Section 5.4, where it is also explained that, in order to achieve more reliable performance measurements, the random sub-sampling step has to be repeated several times. Some of the exercises at the end of this chapter involve the repeated validation sub-sampling procedure for the estimation of the performance of the SVM classifier on the speech-music classification task. In this section, our primary goal is to demonstrate the SVM functions and we will, therefore, defer the discussion on validation until the respective section.
- The last line of the code computes the classification accuracy, i.e. the fraction of the correctly classified test samples. In this example, the accuracy was found to be equal to 0.94. The result may vary due to the randomness that is introduced by the dataset splitting operation.

Note: It is important to note that SVMs cannot directly address multi-class problems; they were initially developed for binary scenarios [39, 55, 56]. This implies that the second argument of the `svmtrain` function can only contain two discrete labels (either character strings or integers). If it is desirable that the SVMs are used in a multi-class task, then one of the methods in Section 5.1.3 needs to be adopted.

Before proceeding with the next classification method, let us first revise the problem of *overfitting* in light of the SVMs. As has already been noted, overfitting occurs when the classification algorithm is overwhelmingly adjusted on the training data. Overfitting can be observed when the classifier achieves high performance rates on the training data but it performs poorly on the testset. In other words, the classifier may not be able to exhibit satisfactory generalization performance.

In the case of the SVMs, high values of the C parameter can lead to overfitting. In order to demonstrate the overfitting phenomenon, we focus again on the speech vs music task. For visualization purposes, we have used only two features, namely, the standard deviation of the ZCR and the mean value of the spectral flux sequence. We have trained an SVM with an RBF-kernel for different values of the C parameter and we present the resulting decision boundaries (lines) in Figure 5.7. It can be observed that as C increases, the

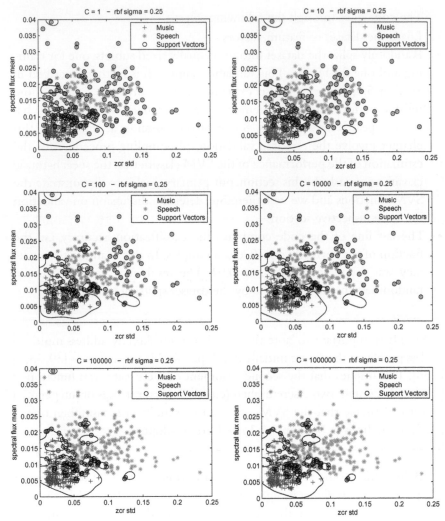

Figure 5.7 SVM training for different values of the C parameter. As C increases, overfitting becomes more evident.

number of correctly classified samples also increases, but the generalization capability of the classifier drops. In Figure 5.8, we present the classification accuracy of this SVM classifier, for different values of C, for the training and the testing data. It can be seen that, for very high C values, the resulting classification scheme manages to classify correctly almost every sample of the training set. However, the classification accuracy on the testset decreases considerably.

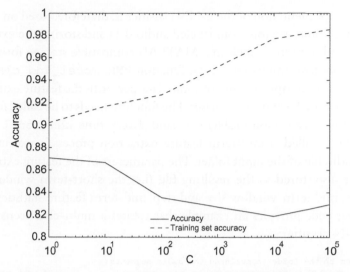

Figure 5.8 Classification accuracy on the training and testing dataset for different values of C. Overfitting occurs at high C values. In those cases, the resulting SVM classifies correctly almost all training samples but fails on a large percentage of the samples of the testing set.

5.3. IMPLEMENTATION-RELATED ISSUES

In this section, we provide instructions related to the library of m–files and datasets that accompany this book. We focus on the proper use of functions that train and test the classifiers and we highlight the most important aspects of the datasets that we provide. This section is essential for readers who want to reproduce the examples of this chapter and reuse the provided code and data in the context of their own development efforts.

5.3.1. Training

As it has already been described, during the training stage the classifier learns the training set, which consists of feature vectors and respective class labels. However, certain classifiers do not adhere to the learning paradigm in the strictest sense. A notable example is the k-NN classifier, which assigns a test sample to a class based on its distance from the instances of the training vectors. Due to its algorithmic behavior, the k-NN classifier can also be studied in the context of instance-based classifiers [57] and, if the training instances are stored in memory, it can be thought of as a memory-based scheme.

Our implementation of the k-NN classifier is, therefore, based on simply extracting feature vectors from labeled audio data and storing the extracted features and respective labels in a MATLAB compatible storage format. To this end, we have implemented the function kNN_model_add_class(), which receives as input a folder of audio data, performs the feature extraction stage and stores the extracted dataset. The function needs to be called for each class of the classification problem at hand. Every time kNN_model_add_class() is called, a mid-term feature extraction process is executed on every audio file of the input folder. The parameters of the feature extraction stage are also stored in the resulting file (i.e. the short-term window size and step, mid-term window size and step, mid-term feature statistics). The following code provides an example that creates a multi-class scenario for the k-NN classifier:

```
% path to the folder containing the audio segments:
strDir = '/media/DISK80_RESE/ResearchData/AUDIO/movieSegments/8-class/';
% mid-term statistics to be used:
Statistics = {'mean','median','std','stdbymean','max','min'};
% short-term and mid-term processing window length and step:
stWin = 0.040; stStep = 0.040;
mtWin = 2.0;    mtStep = 1.0;
% Perform feature extraction and store the extracted features:
kNN_model_add_class('model8.mat', 'music', [ strDir 'Music/'], ...
   Statistics, stWin, stStep, mtWin, mtStep);
kNN_model_add_class('model8.mat', 'speech', [ strDir 'Speech/'], ...
   Statistics, stWin, stStep, mtWin, mtStep);
kNN_model_add_class('model8.mat', 'others1', [ strDir 'Others1/'], ...
   Statistics, stWin, stStep, mtWin, mtStep);
kNN_model_add_class('model8.mat', 'others2', [ strDir 'others2/'], ...
   Statistics, stWin, stStep, mtWin, mtStep);
kNN_model_add_class('model8.mat', 'others3', [ strDir 'others3/'], ...
   Statistics, stWin, stStep, mtWin, mtStep);
kNN_model_add_class('model8.mat', 'shots', [ strDir 'Shots/'],
   Statistics, stWin, stStep, mtWin, mtStep);
kNN_model_add_class('model8.mat', 'fights', [ strDir 'Fights/'],
   Statistics, stWin, stStep, mtWin, mtStep);
kNN_model_add_class('model8.mat', 'screams', [ strDir 'Screams/'],
   Statistics, stWin, stStep, mtWin, mtStep);
```

In this example, the length of the short-term window was set equal to 40 ms (no overlap) and the mid-term window was 2 s long, with a 50% overlap between successive windows. In total, 6 mid-term statistics were extracted per audio file. The code of the example creates a classification setup for the 8-class problem in Section 4.2. The resulting dataset is stored in the model8.mat file.

Table 5.1 Classification Tasks and Files

Name	Description	# Classes	# Samples Per Class
model8	Movie segments (Section 4.2)	8	480
model4	4-class task: music, male speech, female speech, and silence (see also Exercise 1)	4	55
modelSM	Speech vs music	2	480
modelMusicGenres3	Three musical genres	3	120
modelSpeech	Speech vs non-speech	2	480

In a similar manner, we have created the modelMusicGenres3.mat file which addresses a 3-class task for the genres of classical, jazz, and electronic music. In addition, for the binary classification task of *speech vs music*, we have implemented the modelSM.mat file, using the feature extraction parameters of the 8-class model. Finally, we have created a binary speech vs non-speech task. The non-speech class was populated with samples from several classes. *All resulting datasets are included in the MATLAB Audio Analysis Library which serves as an accompaniment to this book.* Table 5.1 provides a summary of the respective files. Note that if you wish to design your own classification setup, you only need to store the audio data of each class in a separate folder and call the kNN_model_add_class() function separately for each folder. All classification tasks, apart from model4, have been evaluated using the *k*-NN classifier and the respective results are given in Section 5.5. The evaluation of model4 is given as an exercise at the end of this chapter.

Note: The mid–term analysis stage may produce a sequence of mid–term feature vectors per audio file, depending on the duration of the file and the length of the mid–term window. However, we only want a single feature vector to represent each signal. Therefore, the kNN_model_add_class() function performs a long–term averaging step over the extracted mid–term feature vectors and finally yields one feature vector per audio file.

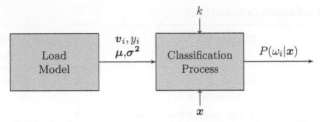

Figure 5.9 Implementation of the *k*-NN classification procedure. At first, the stored data (feature vectors, true class labels and related parameters) are loaded into memory. Then, the classifier assigns the unknown sample to a class based on the *k*-NN algorithm.

5.3.2. Testing

Assuming that the generation and storage of the datasets have been completed, we can proceed to the testing phase. In order to classify an unknown sample, we first load the classification setup into memory (from the corresponding mat-file) and then execute the classification step (Figure 5.9).

We now focus on the *k*–NN classifier to facilitate the presentation of the various implementation issues related to the testing phase. To begin with, the feature vectors of the training set may need to be *normalized* before they are eventually used by the classifier. For example, if the Euclidean distance has been adopted, certain features can dominate the computation of the distance measure due to their range of values. A remedy to this issue is to use the linear method of Eq. (5.14) to normalize the jth feature, $j = 1, \ldots, L$, to zero mean, and standard deviation equal to 1:

$$\widehat{v}_i(j) = \frac{v_i(j) - \mu(j)}{\sigma(j)}, \quad i = 1, \ldots, M, j = 1, \ldots, L, \tag{5.14}$$

where M is the number of training samples, L is the dimensionality of the feature space, $\mu(j)$ is the mean value of the jth feature, and $\sigma(j)$ the respective standard deviation. The normalization parameters (mean value and standard deviation) are not stored in the mat-file of the training set along with the training data; they are computed right after the model is loaded (see below) and are used to normalize the training data and each test sample before it is classified.

The function that loads the required data and computes the normalization parameters (mean value and standard deviation), is `kNN_model_load()`. After the data have been loaded, a test sample can be classified using the `classifyKNN_D_Multi()` function, which implements the *k*-NN classifier for the generic multi–class scenario. The `fileClassification()`

function demonstrates how to classify an unknown audio segment, which is assumed to be available as a WAVE file, using the functionality of the kNN_model_load() and classifyKNN_D_Multi() m-files.

```
function [label, P, classNames] = ...
    fileClassification(wavFileName, kNN, modelFileName)


%
% function [label, P, classNames] = ...
%       fileClassification(wavFileName, kNN, modelFileName)
%
% This function demonstrates the classification of an audio segment,
% stored in a wav file.
%
% ARGUMENTS:
% — wavFileName:     the path of the wav file to be classified
% — kNN:             the k parameter of the kNN algorithm
% — modelFileName:   the path of the kNN classification model
%
% RETURNS:
% — label:           the label of the winner class
% — P:               a vector that contains all estimated probabilities
%                    for each audio class contained in the model
% — classNames:      a cell array that contains the names of the
%                    audio classes of the classification model
%
% NOTE: This function classifies the WHOLE audio file, i.e., we
%       assume that the file contains a homogeneous audio segment.
%       For mid—term classification, please use mtFileClassification().
%

% load classification model:
[Features, classNames, MEAN, STD, Statistics, ...
    stWin, stStep, mtWin, mtStep] = kNN_model_load(modelFileName);

[x, fs] = wavread(wavFileName);         % read wav file
% short—term feature extraction:
stF = stFeatureExtraction(x, fs, stWin, stStep);
mtWinRatio = mtWin / stWin; mtStepRatio =  mtStep / stStep;
% mid—term feature statistic calculation:
[mtFeatures] = mtFeatureExtraction(...
    stF, mtWinRatio, mtStepRatio, Statistics);
% long term averaging of the mid—term statistics:
mtFeatures = mean(mtFeatures,2);
% kNN classification
[P, label] = classifyKNN_D_Multi(Features, ...
    (mtFeatures — MEAN') ./ STD', kNN, 1);
```

Note that in the implementation of the fileClassification() function, the mid-term window size was set equal to the length of the audio segment, since we wanted to extract a single vector of feature statistics from the whole file. For non-homogeneous (and possibly longer) audio streams, a segmentation stage has to be applied before the classification procedure takes place. As will be described later in this book, a simple way to segment an audio signal is to split it into fixed-size segments and classify each segment separately. For the interested reader, the segmentation functionality has been implemented in the `mtFileClassification()` function.

5.4. EVALUATION

The goal of this section is to provide basic tools for the evaluation of the performance of classification methods. We will describe measures that quantify the performance of a classifier on a dataset and we will show how we can partition a dataset into training and testing subsets so that the extracted performance measures are reliable. We will also present performance results for selected case studies in the context of audio segment classification. Although the case studies will be based on the k-NN method, any other type of classifier can also be used.

5.4.1. Performance Measures

An important tool for analyzing the performance of binary and multi-class methods is the *confusion matrix*, which provides the means to group the classification results into a single matrix and helps the designer of the classifier to understand the types of errors that occur during the testing (and training) stage. The confusion matrix, CM, is a $N_c \times N_c$ matrix, whose rows and columns refer to the true (ground truth) and predicted class labels of the dataset, respectively. In other words, each element, $CM(i, j)$, stands for the number of samples of class i that were assigned to class j by the adopted classification method. It follows that the diagonal of the confusion matrix captures the correct classification decisions $(i = j)$.

In many applications, it is useful to normalize the confusion matrix so that its elements become probabilities and not simple counts of events. This can be basically achieved in two ways, the first of which is to divide each element of CM by the total number of samples in the dataset, i.e. by the sum of the elements of the confusion matrix:

$$CM_n(i, j) = \frac{CM(i, j)}{\sum_{m=1}^{N_c} \sum_{n=1}^{N_c} CM(m, n)}. \tag{5.15}$$

According to this type of normalization, the interpretation of $CM_n(1, 2) = 0.12$ is that 12% of the samples of the whole dataset were misclassified to class 2, although they actually belonged to class 1.

A second type of normalization can take place row-wise, if we divide each element of the confusion matrix by the sum of elements of the respective row, i.e. by the (true) population of the class that has been mapped on the row. After normalization has taken place, we can no longer derive from the confusion matrix the count of samples per class, which means we discard the information that is related to the size of each class. In other words, all classes are considered to be of equal size and the dataset to be class-balanced. Obviously, the elements of each row sum to unity.

$$CM_n(i, j) = \frac{CM(i, j)}{\sum_{n=1}^{N_c} CM(i, n)}. \tag{5.16}$$

As an example, the interpretation of $CM_n(1, 2) = 0.12$ is that 12% of the samples *of the first class* (and not of the whole dataset) were misclassified to class 2.

Based on the standard version of the confusion matrix (before normalization takes place), it is possible to extract three useful performance measures, the first of which is the *overall accuracy*, *Acc*, of the classifier, which is defined as the fraction of samples of the dataset that have been correctly classified. It can easily be seen that the overall accuracy can be computed by dividing the sum of the diagonal elements (number of correctly classified samples) by the total sum of the elements of the matrix (total number of samples in the dataset).

$$Acc = \frac{\sum_{m=1}^{N_c} CM(m, m)}{\sum_{m=1}^{N_c} \sum_{n=1}^{N_c} CM(m, n)}. \tag{5.17}$$

Obviously, the quantity $1 - Acc$ is the overall classification error.

Apart from the overall accuracy, which characterizes the classifier as a whole, there also exist two class-specific measures that describe how well the classification algorithm performs on each class. The first of these measures is the *class recall*, $Re(i)$, which is defined as the proportion of data with true class label i that were correctly assigned to class i:

$$Re(i) = \frac{CM(i, i)}{\sum_{m=1}^{N_c} CM(i, m)}, \tag{5.18}$$

where $\sum_{m=1}^{N_c} CM(i, m)$ is the total number of samples that are known to belong to class i. Note, that if the confusion matrix has been row-wise normalized, then $\sum_{m=1}^{N_c} CM(i, m) = 1$, and as a result, $Re(i) = CM(i, i)$, which means that the diagonal elements of the matrix already contain the respective recall values.

The second class-specific performance measure is *class precision*, $Pr(i)$, which is defined as the fraction of samples that were correctly classified to class i if we take into account the total number of samples that were classified to that class. Class precision is, therefore, a measure of accuracy on a class basis and is defined according to the equation:

$$Pr(i) = \frac{CM(i, i)}{\sum_{m=1}^{N_c} CM(m, i)}, \qquad (5.19)$$

where the quantity in the denominator, $\sum_{m=1}^{N_c} CM(m, i)$, stands for the total number of samples that were classified to class i.

It should be noted that, if all classes contain the same number of samples, i.e. if all classes are a *priori* equiprobable, then the above three performance measures can be computed from any version of the confusion matrix, regardless of normalization. On the other hand, if the classes are unbalanced, the second normalization method will produce different performance measures compared with the standard version of the matrix and the first normalization scheme.

If we examine class precision and recall jointly, we can reach the conclusion that there exists a trade-off between these two performance measures and this can be more easily understood if we draw a binary example. To this end, let us assume that we deal with the problem of speech (class ω_1) vs non-speech (class ω_2) audio segments and let each audio segment be represented by a single feature. For simplicity, we assume that this audio feature exhibits a Gaussian (normal) distribution in each class. If $\mu_1 = 0$, $\mu_2 = 3$, $\sigma_1 = 1$, and $\sigma_2 = 1.5$ are the mean values and standard deviations of the feature in the two classes, the respective probability density functions are:

$$P(x|\omega_1) = \mathcal{N}(0, 1),$$

and

$$P(x|\omega_2) = \mathcal{N}(3, 1.5),$$

where

$$\mathcal{N}(\mu, \sigma^2) = \frac{1}{\sqrt{2\pi}\sigma} e^{-\frac{(x-\mu)^2}{2\sigma^2}},$$

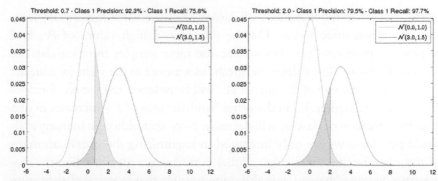

Figure 5.10 Binary classification task with Gaussian feature distributions and two different decision thresholds. If the threshold is shifted to the right, the recall of the first class is increased and its precision is decreased.

is the one-dimensional Gaussian probability density function. If we assume that the two classes are equiprobable (i.e. their prior probabilities, $P(\omega_1)$ and $P(\omega_2)$, are equal), then the corresponding Bayesian classifier becomes equivalent to a thresholding rule, which can be directly applied on the feature value. Specifically, if T is the classification threshold and $x < T$, then the sample is assigned to class ω_1, otherwise it is classified to class ω_2. The optimal threshold is the one that minimizes the classification error and it is naturally derived as the solution of the equation

$$P(x|\omega_1) = P(x|\omega_1),$$

which determines the boundary between the two classes in a Bayesian formulation of the classification task. Obviously, threshold changes affect the performance of the classifier. Figure 5.10 presents two examples based on two different thresholds. In the first case the threshold is set equal to 0.7. The two shaded areas in each figure contribute to the probability of the classification error. More specifically, the shaded region on the left (colored red)[1] contributes to the error probability, Pe_2, that a sample belongs to class ω_2 and was classified to ω_1. Similarly, the area on the right (colored green) refers to the error probability, Pe_1, that a sample belongs to class ω_1 and was misclassified to ω_2.

We are now interested in examining how the recall and precision rates of the first class are affected by the choice of threshold. The higher the value

[1] For interpretation of the references to color in this figure legend, the reader is referred to the web version of this book.

of Pe_1, the lower the recall rate of class ω_1, because more samples from ω_1 class are misclassified to ω_2. On the other hand, high values of Pe_2 lead to low precision rates for class ω_1 because more samples from ω_2 class are misclassified to ω_1. If the decision threshold is moved to the right (to a higher feature value), the size of the area of Pe_1 will be reduced and the recall rate of class ω_1 will be higher. If, on the other hand, the area of Pe_2 increases in size, then the precision of class ω_1 will decrease. Note that, although in many real-world problems, we are only interested in maximizing the overall accuracy of the classifier (or some other global performance measure, like the F_1-measure described below), it is also quite possible that in many classification tasks we want to tune the performance of the classifier so as to achieve higher precision or recall rates for selected classes.

Finally, a widely used performance measure that combines the values of precision and recall is the F_1-*measure*, which is computed as the harmonic mean of the precision and recall values:

$$F_1(i) = \frac{2Re(i)Pr(i)}{Pr(i) + Re(i)} \tag{5.20}$$

We provide the aforementioned performance measures in the MATLAB function `computePerformanceMeasures()`. The function receives as input the estimated class labels and the respective ground truth data.

Note: The Statistics Toolbox of MATLAB also provides a function to compute the confusion matrix given the estimated and true class labels. We provide our own version in an effort to minimize dependency on the functionality of the MATLAB toolboxes and this has been our concern in many similar situations throughout this book.

5.4.2. Validation Methods

A crucial stage in the life cycle of any classifier is the *validation stage*, during which we are asked to verify the correctness of the results that the classifier produces. The validation procedure is usually associated with several decisions that the system designer needs to make, the most important of which has to do with the choice of the training and testing datasets. Remember that we usually have at our disposal a single dataset that consists of feature vectors and respective class labels. Therefore, an important question is how this dataset should be partitioned in order to yield a training and a testing dataset. Once this decision has been made, it is then possible to train the classifier on the training set and measure its performance on the testset using any of the performance measures that were presented in the previous section (overall accuracy, class precision/recall, F_1 measure). The ultimate goal of the validation stage is to help the designer understand if the classifier is capable of generalizing its performance on data outside the training set. In this way, it is also possible to perform a comparative study among classifiers and choose the best one for the task at hand.

When it comes to performance evaluation, the term *cross-validation* is frequently used to indicate that the available dataset is split into two subsets, one for training and one for testing and that the validation of the classifier is achieved by cross-examining these two datasets. Several cross-validation schemes have been proposed over the years for various analysis tasks. Some common schemes are:

- **Resubstitution validation**. This is the simplest way to measure the performance of a classification system. All samples of the dataset are used for both training and testing. Obviously, this method can easily result in overfitting because we do not receive any evidence on the performance of the classifier on unseen data. Equivalently, we cannot be sure whether the classifier is capable of generalization. Although there have been studies that provide certain hints about the reliability of this validation method with respect to the size of the dataset, in general, it is advised that the method is avoided in practice due to the increased risk of overfitting.

- **Hold-out validation**. In an attempt to avoid overfitting, the hold-out method partitions the dataset into two non-overlapping subsets: one for training and the other for testing. For example, one third of the samples are used for testing and the rest for training. From the point of view of the classifier, the testset consists of unknown samples and it is therefore possible to extract a more reliable estimate of the generalization capabilities of the classifier. However, this method is still sensitive to the choice of datasets. For example, important (representative) samples can be left out of the training set, leading to a less accurate classifier. On the other hand, if the number of testing samples is significantly reduced in order increase the size of the training set, then our confidence in the derived performance measures is also likely to decrease.

- **Repeated random sub-sampling validation (hold-out)**. A remedy for the weakness of the standard hold-out method is to repeat the method k times. At each iteration, the dataset is randomly split into two subsets (random sub-sampling). The classification performance is derived by averaging the hold-out iterations. Due to the random nature of the splitting process, it is still possible that some samples may never participate in the training (or the testing) dataset, however, compared to the standard hold-out method the risk of overfitting is reduced.

- **K-fold cross-validation**. The dataset is randomly split into k non-overlapping subsets (the folds), of (approximately) equal size. The classifier is then trained and tested k times: each time a single fold is reserved for testing and the remaining $k - 1$ subsets compose the training dataset. During the splitting procedure in the beginning of the method, care must be taken so that each fold is a good representative of the complete dataset. In order to achieve this goal, we can use a random permutation of the samples before the splitting action takes place.
- **Repeated k-fold cross-validation**. This method runs the k-fold cross-validation approach multiple times, in order to achieve a more reliable estimate of the classification performance. Obviously, a new permutation of the samples needs to occur in the beginning of each iteration.
- **Leave-one-out cross-validation**. The leave-one-out method is actually a variation of the k-fold cross validation approach, where $k = M$, i.e. the number of folds is equal to the total number of samples available in the set. In other words, each fold consists of a single sample. Therefore, during each iteration, all the samples, apart from one, are used for training the classifier and the remaining sample is used in the testing stage. The leave-one-out method is an exhaustive validation technique that can produce very reliable validation results. However, its computational burden can be prohibitive in many applications that involve large datasets.

The software library that accompanies this book includes an implementation of the repeated hold-out and the leave-one-out validation techniques. The name of the corresponding function is `evaluateClassifier()`, which returns the first normalized version of the confusion matrix, the overall accuracy, the recall and precision rates for each class, and the F_1 measure on a class basis. It also returns the row-wise normalized confusion matrix, along with all the respective performance measures. Finally, the accompanying function `printPerformanceMeasures()` prints a table (in LaTeX format) with the confusion matrix and the associated performance measures. The following function (named `scriptClassificationPerformance`) demonstrates how to call these two functions in order to evaluate the performance of the k-NN classifier on a given dataset while varying the number of neighbors, k. In the end, the performance of the best k-NN classifier is printed.

```
function scriptClassificationPerformance(modelName)

% function scriptClassificationPerformance(modelName)
%
% This function:
% A) Loads a kNN audio classification model and computes the
%    respective performance measures using the repeated holdout
%    and the leave-one-out validation methods.
% B) For the best k (for both evaluation methods), it prints the
%    respective confusion matrix and class-specific performance measures
%    (recall, precision and F1 measure).
%
% NOTEs:
% - The optimization is based on the overall accuracy
% - The performance measures are based on the row-wise normalized CMs
%

% load classification model:
load(modelName); ks = [3:2:19]; % possible k values

% TESTING (EVALUATION)
for i=1:length(ks) % for each k value:
    % repeated hold-out validation:
    [CM1A{i}, Ac1A(i), Pr1A{i}, Re1A{i}, F11A{i}, ...
        CM2A{i}, Ac2A(i), Pr2A{i}, Re2A{i}, F12A{i}] = ...
            evaluateClassifier(Features, ks(i), 1, [0.90 20]);
    % leave-one-out:
    [CM1B{i}, Ac1B(i), Pr1B{i}, Re1B{i}, F11B{i}, ...
        CM2B{i}, Ac2B(i), Pr2B{i}, Re2B{i}, F12B{i}] = ...
            evaluateClassifier(Features, ks(i), 2, []);
end
% only CM-row-wise normalized measures are kept
% (second arguments of evaluateClassifier):
F12Amean = mean(cell2mat(F12A(:)), 2); % (average F1 measure)
F12Bmean = mean(cell2mat(F12B(:)), 2);
% plot results
figure;  subplot(2,1,1); hold on;
title('Results - repeated hold-out validation');
plot(ks, Ac2A);  plot(ks, F12Amean, 'r');
xlabel('k'); ylabel('Performance');
legend('Overall Accuracy', 'F1 measure');
subplot(2,1,2); hold on;
title('Results - leave-one-out validation');
plot(ks, Ac2B);  plot(ks, F12Bmean, 'r');
xlabel('k'); ylabel('Performance');
legend('Overall Accuracy', 'F1 measure');
save([modelName '_results']); % save resultss
```

```
% PRINT LATEX CONFUSION MATRIX FOR THE BEST k HERE:
% for the repeated-hold-out validation method:
[MAXA, IMAXA] = max(Ac2A);
fprintf('\n\n * * *    repeated-hold-out (bestK=%d) * * * \n\n', ks(IMAXA));
printPerformanceMeasures(CM2A{IMAXA}, Ac2A(IMAXA), ...
    Pr2A{IMAXA}, Re2A{IMAXA}, F12A{IMAXA}, ClassNames);
% for the leave-one-out validation method:
[MAXB, IMAXB] = max(Ac2B);
fprintf('\n\n * * *    leave-one-out (bestK=%d) * * * \n\n', ks(IMAXB));
printPerformanceMeasures(CM2B{IMAXB}, Ac2B(IMAXB), ...
    Pr2B{IMAXB}, Re2B{IMAXB}, F12B{IMAXB}, ClassNames);
```

5.5. CASE STUDIES

Section 5.3.1 presented the design of a k-NN classifier for presegmented audio data. In this section, we will provide performance measurements and respective comments for some representative case studies. Note that we have chosen to present the results based on row-wise, normalized confusion matrices (see Section 5.4.1) because we wanted to avoid dependence on the a *priori* class probabilities. Furthermore, for the sake of reusability of the designs by the user, we provide recommended k-values for the tasks being studied. As explained in the rest of the section, the recommended values depend on the size of the dataset (i.e. the total number of training samples) and are therefore dataset-specific. If the users decide to use their own datasets for training, certain parameter tuning will be necessary.

5.5.1. Multi-Class Audio Segment Classification

We begin with a case study which presents the performance of the k-NN classifier on an 8-class audio event classification problem. The audio segments for this task are considered to be homogeneous and have been extracted from the audio streams of various movies that cover a wide range of genres. The classes of this problem were described in Section 4.2. The mean duration of each segment is approximately 2 s.

Figure 5.11 presents the overall accuracy and the F1 measure of the k-NN algorithm for different values of the k parameter. It can be seen that the performance of both validation methods (repeated hold-out and leave-one-out) is maximized for $k = 3$. Table 5.2 shows the row-wise normalized confusion matrix and respective performance measures for the leave-one-out method and for $k = 3$.

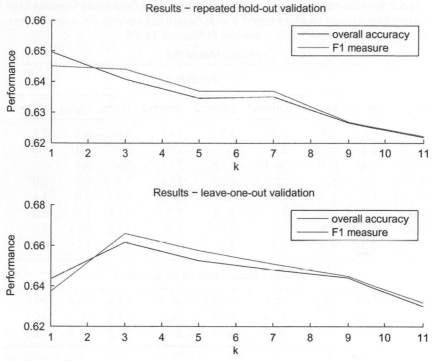

Figure 5.11 Performance of the k-NN classifier on an 8-class task, for different values of the k parameter and for two validation methods (repeated hold-out and leave-one-out). The performance measures were computed based on the row-wise normalized confusion matrix. The best performance ($F1$ measure) is achieved when $k = 3$ for both validation methods.

5.5.2. Speech vs Music Classification of Audio Segments

Speech-music discrimination can be a useful preprocessing stage in several multimedia systems, including automatic monitoring of radio broadcasts, speech recognition, low-bit rate audio coding, and generic audio segmentation. The term refers to the problem of splitting an audio stream into homogeneous segments and classifying each segment as speech or music. A third class is often used to indicate background noise and/or silence. Over the years, several solutions to this problem have been proposed, e.g. [24, 16]. In this case study we ignore the segmentation stage, assume that homogeneous audio excerpts are available and focus on the simpler binary classification task of speech vs music for presegmented data [18, 58–60]. Table 5.3 presents the classification performance of the k-NN classifier for this binary task. It can be seen that the overall accuracy exceeds 96% for the dataset being studied.

Table 5.2 Row-Wise Normalized Confusion Matrix for the 8-Class Audio Segment Classification Task (Leave-One-Out Method, $k = 3$). Overall Accuracy: 66.2%, Average Precision: 68.4%, Average Recall: 66.2%, Average F1 Measure: 66.6%

Confusion Matrix (%)

				Predicted				
True ⇓	Music	Speech	Others1	Others2	Others3	Shots	Fights	Screams
Music	64.2	1.7	6.7	3.5	13.1	5.6	4.4	0.8
Speech	1.2	89.6	0.2	2.5	1.9	0.4	3.5	0.6
Others1	2.8	0.7	56.4	13.2	21.6	2.4	2.1	0.7
Others2	3.0	1.5	12.9	60.7	13.4	7.0	1.5	0.0
Others3	6.6	1.3	15.8	3.3	59.2	8.6	2.0	3.3
Shots	0.6	0.2	2.1	10.8	7.7	77.1	1.2	0.2
Fights	1.7	3.5	1.0	12.1	7.3	13.3	60.4	0.6
Screams	3.3	5.4	3.1	1.7	9.8	4.4	10.6	61.7

Performance Measurements (%, per class)

Precision	76.9	86.2	57.4	56.3	44.2	64.9	70.4	90.8
Recall	64.2	89.6	56.4	60.7	59.2	77.1	60.4	61.7
F1	70.0	87.9	56.9	58.4	50.6	70.5	65.0	73.4

Table 5.3 Row-Wise Normalized Confusion Matrix for the Speech vs Music Binary Classification Task (Leave-One-Out Method, $k = 3$). Overall Accuracy: 96.2, Average Precision: 96.3, Average Recall: 96.2, Average F1 Measure: 96.2 (numbers in %)

Confusion Matrix (%)

	Predicted	
True ⇓	Speech	Music
Speech	97.7	2.3
Music	5.2	94.8

Performance Measurements (per class)

Precision	94.9	97.6
Recall	97.7	94.8
F1	96.3	96.2

5.5.3. Musical Genre Classification

The development of computational tools for the task of automatic musical genre classification has been an applied research topic for over a decade, starting from the early work [14] that laid the foundations for a systematic treatment of this particular classification task. Musical genre classification can be considered an ill–defined problem due to the inherent difficulty in establishing general definitions of different musical genres. However, research has shown that it is possible to develop classification systems that exhibit satisfactory performance for certain genres of popular music, as is the case with the musical genres in the GTZAN dataset [15].

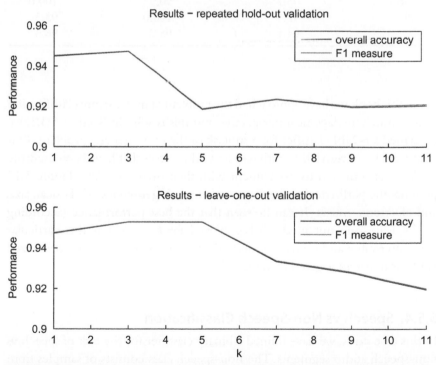

Figure 5.12 Estimated performance for the 3-class musical genre classification task, for different values of the k parameter and for two evaluation methods (repeated hold-out and leave-one-out). The performance was evaluated on the row-wise normalized confusion matrix. The best performance was achieved for $k = 3$, according to both validation methods.

Table 5.4 Row-Wise Normalized Confusion Matrix for the 3-Class Musical Genre Classification Task (Leave-One-Out Evaluation Method, $k = 3$). Overall Accuracy: 95.3, Average Precision: 95.5, Average Recall: 95.3, Average F1 Measure: 95.3 (Numbers in %)

	Confusion Matrix (%)		
	Predicted		
True ⇓	Classical	Jazz	Electronic
Classical	92.5	7.5	0.0
Jazz	0.0	100.0	0.0
Electronic	4.2	2.5	93.3
	Performance Measurements (per class)		
Precision	95.7	90.9	100.0
Recall	92.5	100.0	93.3
F1	94.1	95.2	96.6

In this book we have approached the problem of music genre classification from a purely pedagogical perspective and this is why, in Section 5.3.1, we described a k-NN classifier for a subtask of the more general problem. Our treatment is not restrictive with respect to the classes that have been used; the reader is encouraged to experiment with their own class data. Figure 5.12 presents the performance measurements for a 3-genre task of classical, jazz, and electronic music. It can be seen that the best performance (according to the leave-one-out method) is achieved for $k = 3$. For this particular parameter value and validation method, the respective row-wise normalized confusion matrix is given in Table 5.4.

5.5.4. Speech vs Non-Speech Classification

In this case study, we have trained a binary classifier for the task of speech vs non-speech audio segments. The non-speech class consists of samples from several different audio classes, including environmental sounds, music, etc. This type of classifier can be used as a subsystem in *speaker diarization*, and *speaker verification* applications, in order to filter out the parts of the audio stream that do not contain the speech signal. *Speaker diarization* [61] refers to the problem of answering the question 'who spoke when?', given an audio stream. In *speaker verification* [62], the speech signal is verified against

Table 5.5 Row-Wise Normalized Confusion Matrix for the Speech vs Non-Speech Classification Task (Leave-One-Out Validation Method, $k = 11$). Overall Accuracy: 94.8, Average Precision: 94.9, Average Recall: 94.8, Average F1 Measure: 94.8 (Numbers in %)

Confusion Matrix (%)		
	Predicted	
True ⇓	**Speech**	**Non-speech**
speech	97.1	2.9
non-speech	7.5	92.5
Performance Measurements (per class)		
Precision	92.8	96.9
Recall	97.1	92.5
F1	94.9	94.7

a presented speaker identity, as is, for example, the case with systems that control the entrance to restricted areas.

Table 5.5 demonstrates the row-wise normalized confusion matrix for the current case study, where the leave-one-out method was adopted for validation. In this case, the recommended value for k is 11 and this can be perceived as an indication of a harder classification task compared to the binary problem of speech vs music.

Overall comments on the case studies

- It can be readily observed that in the above case studies the class-specific performance measures (recall, precision, and the F1 measure) may vary among the audio classes. For example, in the 8-class audio event classification task, the 'speech' class is associated with a high accuracy measurement (around 90%), while the performance for the class of environmental sounds is significantly lower.
- The intensity of confusion between two classes, i and j, can be better understood if we observe the quantities $CM(i, j)$ and $CM(j, i)$, where CM is the row-wise normalized version of the confusion matrix. $CM(i, j)$ is the percentage of data of class i that are misclassified to class j and $CM(j, i)$ is the opposite. For the row-wise normalized confusion matrix the arithmetic average of these two quantities can be interpreted as the joint 'confusion' between the two classes.

- The measured performance in our simple case study on musical genres was rather high for reasons that have already been explained. Note, that in practice, the number of genres can be high and/or the class taxonomy can be quite confusing. For example, discriminating between genres of dance music (e.g. trance, house, techno, trip-hop, etc.) can be, in general, a harder task compared to a situation where jazz and electronic music are the only two classes being studied. As a final remark, the quality of annotations of music collections has also been a matter of research and certain studies have implied that disambiguation is necessary before the actual validation process begins, e.g. [63].

5.6. EXERCISES

1. (D1) Use `model4.mat` to evaluate the classification performance of the k-NN classifier. Report the confusion matrix and the respective recall, precision, and F1 measures for each class. *Hint: Use* `scriptClassificationPerformance()`.

2. (D3) Load the `model8.mat` file that contains the feature vectors of the 8-class audio classification task of the respective case study. Extract from this file the data of classes 6 and 7 (gunshots and fights). Use the leave-one-out method to determine the value of parameter k of the k-NN classifier that results in the best classification performance for this binary task (fights vs gunshots). For this best value of k, print the confusion matrix along with the respective performance measures.

3. (D3) Write a MATLAB function that takes as input a) a three-dimensional training set, and b) a vector of weights that define a plane. The output of the function is the accumulated cost of all classification failures with respect to the input plane. Test your function as follows: load the dataset that refers to the speech-music task and select three features that represent mid-term statistics. Split the resulting dataset into two equally sized subsets and apply your function on one of the two resulting subsets. In other words, you are asked to implement Eq. (5.12) for the case of three features. Finally, plot the dataset along with the resulting line.

4. (D4) Implement the perceptron learning algorithm as a MATLAB function. The input to the function is the training dataset, the respective class labels, and the ρ parameter (constant). The dimensionality of the feature space will be inferred from the training set and the m-file should be able to operate on any number of features. The output is the vector of weights that define the hyperplane that separates the two classes. Insert a

mechanism that terminates the training procedure if a maximum number of epochs has been reached, so that you can deal with non-separable classes. To test your function, load the model8.mat file and select a binary problem. Apply the leave-one-out method to validate the classifier for the case of 3, 4, and 5 features of your choice. In the experimental setup, replace your perceptron implementation with MATLAB's newp function. Do you obtain similar results?

5. (D2) Generate and plot a decision tree (like the one in Figure 5.6) trained on the feature space of the 3-class musical genre classification task that is stored in the modelMusicGenres3.mat file.

6. (D4) Create a MATLAB function that performs the following tasks:

   (a) Loads the model8.mat file and generates all possible binary classification tasks (e.g. music vs speech, speech vs others1, etc.).

   (b) For each binary task, estimates the best k value for the k-NN classifier and computes the performance measures for this value. To this end, the leave-one-out validation method is used in conjunction with the row-wise normalized confusion matrix.

   (c) Determines the three 'hardest' binary tasks, i.e. the three tasks that exhibit the highest error rates. The error rate is computed as $1 - Ac$, where Ac is the *classification accuracy* at each task.

   Are the results in accordance with the interpretation of the overall confusion matrix of the 8-class task (Table 5.2)?

7. (D4) Section 5.2.4 explained how to use certain MATLAB functions to train and test an SVM classifier. In that section, we provided the code to evaluate an SVM with linear kernel on the feature space of the binary classification task of speech vs music. Specifically, one single hold-out validation was used: half of the data were used for training and the rest for testing. Expand the experimental setup by completing the following tasks:

   (a) Use repeated hold-out validation in order to obtain a more accurate measurement of the classification performance.

   (b) Further to the default (linear kernel), evaluate the following kernel types: (a) quadratic, (b) 3-order polynomial, and (c) RBF. Experiment with different values of C, and in the case of the RBF kernel, with different values of σ^2.

   After the experiments have been carried out, answer the following questions:

   (a) Which value of C (and σ^2 for the RBF kernel) results in the best performance for each one of the three methods?

(b) In the case of the RBF kernel, are the selected values for C and σ^2 close to the ones that were presented in Section 5.2.4, for the two-dimensional feature space (Figures. 5.7 and 5.8)?

(c) Which SVM classifier yields the best performance?

(d) Is the best SVM classifier better than the k-NN classifier for the same task (speech vs music), if the results in Section 5.5.2 are taken into account?

8. (D3) Function `kNN_model_add_class()` (Section 5.3.1) is used to add a new audio class to the existing ones in a k-NN-related mat-file. As has already been described, this function performs the feature extraction for the samples of the given class and finally stores them. The simplicity of the k-NN algorithm allows us to add new classes to the dataset without the need for any further processing steps like re-training a classifier model. Remember that the k-NN algorithm uses directly the feature vectors and simply computes distances. Does this procedure apply (a) to the SVM classifiers, and (b) to the decision trees? Justify your answers.

9. (D4) Use all the feature sets of Section 5.5 for training and evaluating a decision tree classifier. Generate confusion matrices and compute the respective performance measures (class precision and recall) for each classification task. Compare the performance of the decision tree to the k-NN scheme on a task-by-task basis.

10. (D5) As described in Section 5.2.4, the SVM classifiers are not tailored to the needs of multi-class problems. In this exercise you are asked to:

(a) Write a MATLAB function that implements the one-vs-all (OVA) method (Section 5.1.3). Then, use the function to extend the SVM operation in a multi-class context.

(b) Repeat with the one-vs-one (OVO) method (Section 5.1.3).

(c) Evaluate and compare the two resulting multi-class schemes for the 3-class musical genre classification task (features in `modelMusicGenres3.mat`).

(d) Compare with the performance of the k-NN classifier, presented in Table 5.4 (when the leave-one-out evaluation is used).

11. (D5) The Bayesian classifier, although optimal with respect to the probability of error, is not really useful in practice due to the need to have precise knowledge of the pdfs in possibly high-dimensional feature spaces. However, in order to design working solutions, it is often desirable to resort to certain assumptions that simplify the classification

problem, even though they only hold partially on the dataset. A frequent assumption is the statistical independence of features, which is often coupled with the assumption that each feature follows a Gaussian (or some other simple parametric) distribution per class. If these two assumptions are adopted, it is possible to design a naive classifier that operates on the Bayesian philosophy. This classifier is also known as the Naive Bayes classifier. Its operation can be summarized in the following steps, assuming a binary classification task and a Gaussian pdf for each feature in each class:

(a) Training: For each feature, $x_i, i = 1\ldots, D$, estimate the mean value and the standard deviation of the respective Gaussian pdf, $P(x_i|\omega_i), i = 1, 2$.

(b) Training: Use the following equation to estimate the joint pdf, $P(\underline{x}|\omega_i)$, of all features, per class:

$$\widehat{P}(\underline{x}|\omega_i) = \prod_{k-1}^{D} P(x_i|\omega_i), i = 1, 2 \qquad (5.21)$$

where $\underline{x} = [x_1, x_2, \ldots, x_D]^T$.

(c) Training: Estimate the a *priori* class probabilities, $p(\omega_1)$ and $p(\omega_2)$. This can be done by dividing the number of samples (size) of each class by the sum of the sizes of the two classes.

(d) Classification: The unknown sample, $\underline{x}_u$, is classified to ω_1 if

$$p(\omega_1) \cdot \widehat{P}(\underline{x}_u|\omega_1) > p(\omega_2) \cdot \widehat{P}(\underline{x}_u|\omega_2)$$

otherwise, it is classified to ω_2.

Implement the Naive Bayes classifier as a MATLAB function. Load the modelMusicGenres3.mat file, select two genres, and apply the Naive Bayes classifier on a five-dimensional feature space of your choice. Use the repeated hold-out validation method and print the performance measurements.

Note that MATLAB provides an object-oriented approach to the Naive Bayes classifier in the Statistics Toolbox. In particular, the NaiveBayes class can be used to train and test a Naive Bayes classifier. It is left as a further exercise for the reader to compare their implementation of this exercise with MATLAB's version of the classifier.

CHAPTER 6

Audio Segmentation

Contents

Segmentation is a processing stage that is of vital importance for the majority of audio analysis applications. The goal is to split an uninterrupted audio signal into segments of homogeneous content. Due to the fact that the term 'homogeneous' can be defined at various abstraction layers depending on the application, there exists an inherent difficulty in providing a global definition for the concept of segmentation.

For example, if our aim is to break a radio broadcast of classical music into homogeneous segments, we would expect that, in the end, each segment will contain either speech or classical music (or silence). In this case, the definition of the term homogeneous is based on the audio types that we expect to encounter in the audio stream and we demand that each segment at the output of the segmentation stage consists of only one audio type [24]. Obviously, a different definition is necessary, when we know that we are processing a speech stream by means of a diarization system which aims at determining 'who spoke when?' [64]. In this case, the definition of homo-geneity is tied with the identity of the speaker, i.e. we can demand that each segment contains the voice of a single speaker. As a third example, consider a music signal of a single wind instrument (e.g. a clarinet). The output of the segmentation stage can be the sequence of individual notes (and pauses) in the music stream. At a more abstract level, we might need to develop a segmentation algorithm, which can break the audio stream of a movie into audio scenes, depending on whether violent audio content is present or not [20].

In this chapter, we first focus on segmentation methods that exploit prior knowledge of the audio types which are involved in the audio stream. These methods are described in Section 6.1 and embed classification decisions in the segmentation algorithm. Section 6.2 describes methods that do not make use of prior knowledge. The methods of this second category either detect stationarity changes in the content of the signal or use unsupervised clustering techniques to divide the signal into segments, so that segments of the same audio type are clustered together.

6.1. SEGMENTATION WITH EMBEDDED CLASSIFICATION

Several algorithms evolve around the idea of executing the steps of segmentation and classification, jointly [60,23,24]. A simple approach is to split the audio stream into fixed-size segments, classify each segment separately, and merge successive segments of the same audio type at a post-processing stage. Alternatively, if fixed-size splitting is abandoned and the segmentation procedure becomes dynamic, the classifier can be embedded in the segmentation decisions [24]. In other words, the endpoints of the segments can be dynamically determined by taking into account the potential decision of the classifier on the candidate segments. In both cases, the output of the segmentation stage is a sequence of (*classified*) homogeneous audio segments. The nature of these methods demands that a trained classifier is available, which can be a major restriction depending on the application context. Unfortunately, a trained classifier can often be a luxury because the training data may not be available or the audio data types may not be fixed in advance. These demands call for an 'unsupervised' approach to segmentation, as will be explained in Section 6.2.

6.1.1. Fixed-Window Segmentation

Fixed-window segmentation is a straightforward approach to segmenting an audio stream if a trained classifier is available. The signal is first divided into (possibly) overlapping mid-term segments (e.g. 1 s long) and each segment is classified separately to a predefined set of N_c audio classes. As a result, a sequence of hard classifications decisions (audio labels), $C_i, i = 1, \ldots, N_{mt}$ is generated at the output of this first stage, where N_{mt} is the number of mid–term segments. The extracted sequence of audio labels is then smoothed at a second (post-processing) stage, yielding the final segmentation output, i.e. a sequence of pairs of endpoints (one pair per detected segment) along with the respective audio (class) labels (Figure 6.2).

Note that, instead of hard classification decisions, the output of the first stage can be a set of soft outputs, $P_i(j) \equiv P(i,j), j = 1, \ldots, N_c$, for the ith segment. The soft outputs can be interpreted as the estimates of the posterior probabilities of the classes, $P(\omega_j \mid \underline{x}_i)$, given the feature vector, $\underline{x}_i$, that represents the mid-term segment (obviously, $\sum_{j=1}^{N_c} P_i(j) = 1$). Therefore, depending on the output of the first stage, we can broadly classify the techniques of the second stage in two categories:

* *Naive merging* techniques: the key idea is that if successive segments share the same class label, then they can be merged to form a single segment (Section 6.1.1.1).
* *Probability smoothing* techniques: if the adopted classification scheme has a soft output, then it is possible to apply more sophisticated smoothing techniques on the sequence that has been generated at the output of the first stage. Section 6.1.1.2 explains in detail a smoothing technique that is capable of processing a sequence of soft outputs and presents its application on an audio stream.

Figure 6.1 provides a schematic description of the post-processing stage. Irrespective of the selected post-segmentation technique, the output is a sequence of pairs of endpoints, $S_i \equiv (S_i(1), S_i(2)), i = 1, \ldots, N_s$ and respective class labels, $L_i, i = 1, \ldots, N_s$, where N_s is the final number of segments. Note that, throughout this chapter we will use the terms 'post-processing,' 'post-segmentation,' and 'smoothing' interchangeably when we refer to the second processing stage of the segmenter.

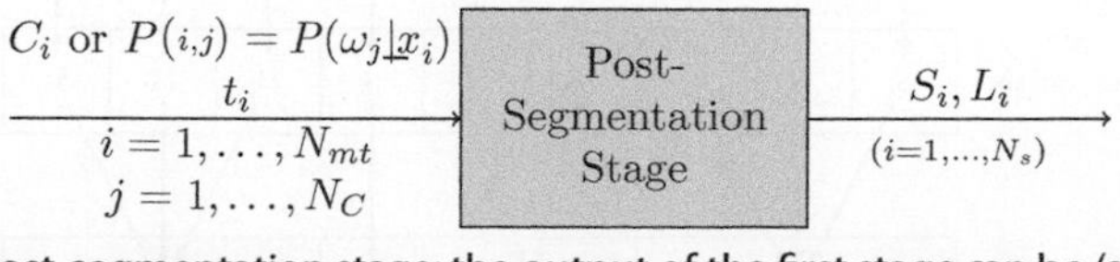

Figure 6.1 Post-segmentation stage: the output of the first stage can be (a) a sequence of hard classification decisions, $C_i, i = 1, \ldots, N_{mt}$; or (b) a sequence of sets of posterior probability estimates, $P_i(j), i = 1, \ldots, N_{mt}, j = 1, \ldots, N_c$. The t_i s provide temporal information, i.e. refer to the center of each fixed-term segment. The output is a sequence of pairs of endpoints, $S_i \equiv (S_i(1), S_i(2)), i = 1, \ldots, N_s$ and respective class labels $L_i, i = 1, \ldots, N_s$, where N_s is the final number of segments.

6.1.1.1. Naive Merging

This simple procedure assumes that a sequence of audio labels (hard classification decisions) has been generated by the first stage of the segmenter. The only action that is performed by the post-segmentation stage is to merge successive mid-term segments of the same class label in to a single, longer segment. This rationale is illustrated in Figure 6.2, where a binary classification problem of speech vs music has been assumed.

The post-processing functionality of the naive merging scheme has been implemented in the `segmentationProbSeq()` m-file of the MATLAB toolbox provided with the book. Note that this function needs to be called after the `mtFileClassification()` function, which generates a set of estimates of the posterior class probabilities for each mid-term window. At each segment, function `segmentationProbSeq()` determines the class label based on the maximum probability and then proceeds with the merging procedure. The first input argument of `segmentationProbSeq()` is a matrix, P, where $P(i, j)$ corresponds to the estimated probability that the ith

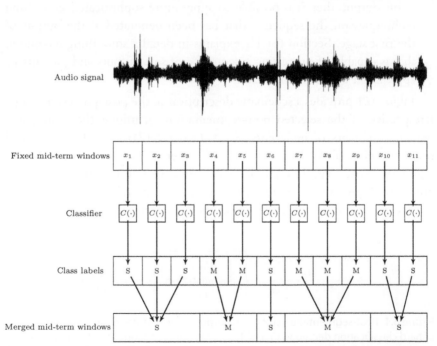

Figure 6.2 Fixed-window segmentation. Each subsequence whose elements exhibit the same hard label merge forming a single segment. 'S' and 'M' stand for the binary problem of speech vs music.

window belongs to the jth class. If the adopted classifier can only provide a hard decision, C_i, for the ith mid-term window, it is still easy to generate the input matrix, P, using the following rule (given in pseudocode):

```
for the i-th mid-term window, i = 1,...,N_mt
    for the j-th class, j = 1,...,N_c
        if C_i equals j then
            P(i,j) = 1,
        else
            P(i,j) = 0
```

In more detail, function `segmentationProbSeq()` accepts the following arguments:

1. The matrix of probabilities, P, which was described above. It has N_{mt} rows (equal to the number of fixed mid-term windows) and N_C columns (equal to the number of classes). Note that in the naive merging mode, only hard decisions are needed, so the `segmentationProbSeq()` function can simply select from each row the class that corresponds to the maximum probability. Several of the examples in this book use the k-NN classifier, which can provide soft outputs. Of course, there is always the option to use the aforementioned rule for the conversion of the output of any classifier that provides hard decisions to a compatible P-matrix.
2. $t_i, i = 1, \ldots, N_{mt}$: the centers (in seconds) of the corresponding mid-term windows.
3. The total duration of the input signal, which is used to generate the endpoints of the last segment. The example at the end of this section demonstrates how to obtain this quantity.
4. The type of post-segmentation technique: 0 for naive merging and 1 for probability smoothing (see Section 6.1.1.2).

The output of the `segmentationProbSeq()` function is a $N_{mt} \times 2$ matrix, S, whose rows correspond to the endpoints of the detected segments, i.e. $S(i, 1)$ marks the beginning of the ith segment and $S(i, 2)$ its end. Both quantities are provided in seconds. Furthermore, `segmentationProbSeq()` returns a vector, L, with the corresponding audio labels. The following code demonstrates how to use the chain of functions `mtFileClassification()` and `segmentationProbSeq()` in the context of fixed-window segmentation with a naive post-processing stage. We focus on the binary task of *speech-music* classification and load the dataset that is stored in the `modelSM.mat` file.

```
% Speech — music discrimination demo:
% Step 1: mid—term classification (classifier trained on speech — music):
[labels, Ps, Conf, centers, classNames] = mtFileClassification(...
    '../data/speech_music_sample.wav', 3, 'modelSM.mat');
% Step 2: extract segment boundaries:
[a, fs] = wavread('../data/speech_music_sample.wav', 'size');
totalDuration = a(1) / fs; % determine the input file's duration in seconds
[segs, classes, Labels] = segmentationProbSeq(...
    Ps, centers, totalDuration, 0); % segmentation (naive merging)
% write results on output:
for i=1:size(segs, 1)
    fprintf('segment %5d of %5d (%6.2f — %6.2f) : %10s\n', ...
        i, size(segs, 1), segs(i, 1), segs(i, 2), classNames{classes(i)+1});
    % play sound:
    [x, fs] = wavread('../data/speech_music_sample.wav', ...
        round([segs(i,1) * fs + 1,  segs(i,2) * fs]));
    soundOS(x, fs);
end
```

The reason that function `segmentationProbSeq()` operates on a matrix of probabilities instead of a sequence of hard decisions, is that we want to provide a single m-file for both post-processing techniques. Therefore, as the reader understands, this is just a software engineering issue and all that is needed for the naive merger to operate is an intermediate conversion of matrix P to hard labels, as explained above.

> *Note:* Before we proceed with the next post-segmentation technique that uses probability smoothing, we present a simple plotting utility that we have implemented in the MATLAB function `segmentationPlotResults()`. It takes as input a sequence of segments and respective class labels following the format of the output arguments of function `segmentationProbSeq()` and plots a colored illustration of the output of the segmenter. The user is able to select segments and listen to their content. The function can be used for visualization purposes irrespective of the segmenter that has been adopted. A related example is presented later in this chapter, in Section 6.2.2.2, in the context of speaker diarization. Complementary functionality is additionally provided by the `segmentationCompareResults()` function, which serves to compare the segmentation results of two methods.

6.1.1.2. Probability Smoothing

The post-segmentation scheme of the previous section merges successive mid-term windows of the same class label and does not really make use of the soft (probabilistic) output of the classifier, whenever such output is available. On the other hand, if the classification probabilities are exploited, a more sophisticated post-processing scheme can be designed. Consider, for example, a binary classifier of speech vs music embedded in a segmenter and assume that for a mid-term window the classification decision is 'marginal' with respect to the estimated probabilities. For instance, if the k-NN classifier is employed with $k = 9$, it could be the case that five neighbors belong to the speech class and the rest of them to the music class. The respective estimates of the posterior probabilities are $\frac{5}{9}$ and $\frac{4}{9}$, respectively. Although the segment is classified to the speech class, there exists a high probability that a classification error has occurred. In other words, the naive merger of the previous section may have to operate on a noisy sequence of classification decisions. In such situations, the performance of the post-processing stage can be boosted if a smoothing technique is applied on the soft outputs of the classifier.

Figure 6.3 is an example of the application of two post-segmentation procedures on the binary classification problem of speech vs music. Figure 6.3a presents the posterior probability estimates for the two classes (speech and music) over time. Figure 6.3b presents the results of the naive merging scheme, and Figure 6.3c refers to the output of a Viterbi-based smoothing technique that will be explained later in the section. The example highlights the fact that the classifier produces marginal decisions near the 20th second of the audio stream because the posterior probabilities of the two classes are practically equal. The simple merging technique treats the respective segment as music. On the other hand, the smoothing approach takes into

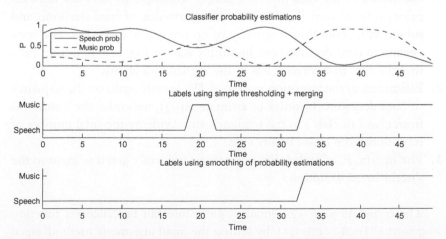

Figure 6.3 Fixed-window segmentation: naive merging vs Viterbi-based smoothing.

account the neighboring probabilistic output of the classifier and generates a 'smoother' decision as far as the transition among class labels is concerned.

The Viterbi algorithm [65, 66] is a dynamic programming algorithm, which, broadly speaking, is aimed at finding the most likely sequence of states, given a sequence of observations. In the context of segmentation, a Viterbi-based algorithm can be designed to determine the most likely sequence of class labels, given the posterior probability estimations on a segment basis [64, 67]. Further to the given probabilities, the Viterbi-based post-segmentation algorithm also needs as input the prior class probabilities and a transition matrix, A, whose elements, $A(i, j)$, $i = 1, \ldots, N_c, j = 1, \ldots, N_c$, indicate the probability of transition from state i to state j, i.e. in our case, from class label i to class label j (assuming N_c classes in total). The core Viterbi algorithm will be further explained in Chapter 7 because it is also an essential algorithm in the context of temporal modeling and recognition. In this chapter, the key idea of the algorithm is preserved and adopted for the purpose of post-processing.

Our version of the Viterbi-based post-processing scheme is tailored to the needs of probability smoothing and it is implemented in function `viterbiBestPath()`. In order to embed the Viterbi-based smoothing procedure in the segmenter, the following initialization steps need to be executed, given a matrix, P, of posterior probability estimations, where $P(i, j)$ corresponds to the estimated probability that the ith segment belongs to the jth class:

1. **Estimation of the class priors.** A simple technique to initialize the class priors is to convert the P matrix to a sequence of hard decisions and subsequently count the frequency of occurrence of each class. The conversion to hard decisions can be simply achieved by selecting from each row the class that corresponds to the highest probability.
2. **Estimation of the transition matrix, A.** We operate again on the sequence of hard decisions. In order to estimate $A(i, j)$, we count the transitions from class i to class j in the sequence and divide by the total number of transitions from class i to any class.
3. **The matrix, P, of probability estimations** is directly given as input to the Viterbi-based technique.

The complete post-segmentation procedure can be called via function `segmentationProbSeq()` by setting the input argument 'method' equal to 1. This is presented in the following piece of code:

```
% estimate the initial transition matrix:
[MAXPs, HardLabels] = max(Probabilities, [], 2);    % get the hard labels
for i=1:numOfClasses    % estimate the class priors:
    priors(i) = length(find(HardLabels==i)) / numOfWindows;
end

transMatrix = zeros(numOfClasses);
for i=2:numOfWindows
  transMatrix(HardLabels(i-1), HardLabels(i)) = ...
      transMatrix(HardLabels(i-1), HardLabels(i)) + 1;
end
transMatrix = transMatrix / sum(sum(transMatrix));
% run the Viterbi-based smoothing technique
Labels = viterbiBestPath(priors, transMatrix, Probabilities');
```

In the end of the above initialization procedure, the 'labels' vector contains the sequence of class labels at the output of the Viterbi algorithm. In other words, this vector contains the sequence of class labels that maximizes the Viterbi score given the estimated (observed) class probabilities on a mid-term basis. After the best-class sequence has been generated, the naive merger is applied on the results in order to yield the final sequence of segments.

6.1.1.3. Example: Speech-Silence Segmentation

The ability to detect the useful parts of an audio stream can be of high importance during the initial processing stages of an audio analysis system. In most applications, the term 'useful' is used to discriminate between the parts of the audio stream that contain the audio types that we are interested in studying and the parts that contain background noise or very low intensity signals. In the rest of this section, we will use the term 'silence' to refer to the latter. In the context of speech processing, the goal of the segmenter is to split the audio stream into homogeneous segments consisting of either speech or silence. The importance of this functionality was already evident in the early speech recognition systems, where it was important to detect the endpoints of isolated words and filter out the silent parts. Silence detection belongs to the broader category of segmentation methods, which demand for prior knowledge of the audio types because a classifier is embedded in the segmenter.

One of the first methods for the detection of the endpoints of a single speech utterance in an audio stream can be found in [8]. The method extracts two time-domain features from the audio signal, namely, the short-term energy and the short-term zero-crossing rate (Section 4.3) and operates on a double thresholding technique. The thresholds are computed based on the assumption that the beginning of the signal does not contain speech. Function `silenceDetectorUtterance()` provides an implementation of this method for the sake of comparison with the other methods that we present in this book. Before you call this silence detector, you need to decide on the short-term frame length and hop size. In the following example, we use overlapping frames (50% overlap) and a frame length equal to 20 ms. The code listed below plots the result, i.e. the endpoints of the extracted segment of speech.

```
% Rabiner's silence detector:
[B1, E1, B2, E2, Er, Zr] = silenceDetectorUtterance(...
     '../data/1WORD.WAV', 0.020, 0.010);
% plot results:
[x, fs] = wavread('../data/1WORD.WAV'); T = 0:1/fs:(length(x)-1)/fs;
plot(T, x); xlabel('Time (sec)'); hold on;
line([B2 B2], [-1 1], 'Color', [1 0 0], 'LineWidth', 2);
line([E2 E2], [-1 1], 'Color', [1 0 0], 'LineWidth', 2);
```

The endpoints of the utterance are stored in variables $B2$ and $E2$. Variables $B1$ and $E1$ contain initial (rough) estimates of the endpoints, which are returned after the first thresholding stage has finished (for more details the reader is prompted to read the description in [8]). Er and Zr are the feature sequences of short-term energy and zero-crossing rate, respectively. In Figure 6.4, we present an example of the application of this approach on an audio signal that contains a single speech utterance.

This silence detector cannot be directly used in the general case of uninterrupted speech, so, in the current example, we go one step further: instead of

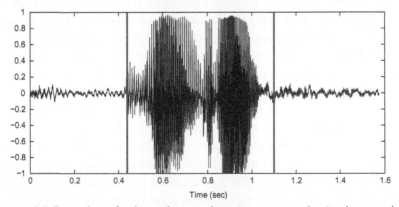

Figure 6.4 Example of the silence detection approach implemented in `silenceDetectorUtterance()`. The detected speech segment's endpoints are shown in red lines (online version) or thick black vertical lines (print version). This approach can only be applied on a single utterance and is not suitable for the general case of uninterrupted speech (at least not after certain re-engineering has taken place).

using thresholding criteria on selected feature sequences, we resort to a *semi-supervised* technique. More details on clustering and semi-supervised learning can be found in Section 6.2.2. In this case study, the key idea is that instead of using an external training set, we generate training data from the signal itself by means of a simple procedure. The method executes the following steps:

1. Given an input signal (e.g. a signal stored in a WAVE file), use function `featureExtractionFile()` to split the signal into mid-term segments and extract all available statistics from each mid-term segment.
2. Focus on the statistic of mean energy: Let S_{ME} be the sorted values of this statistic over all mid-term segments. Store the 20% of lowest values in vector S_{ME1} and, similarly, the 20% of highest values in S_{ME2}.
3. Assume that the values in S_{ME1} correspond to silent segments (low mean energy) and that the values in S_{ME2} originate from speech segments (high mean energy). This assumption can help us generate a training dataset: from each mid-term segment that stems from S_{ME1}, the respective feature vector that consists of all the available mid-term statistics is preserved. The set of these vectors comprises the training data of the class that represents silence. Similarly, the feature vectors that originate from S_{ME2} form the training data of the speech class.
4. The k-NN setup for this binary classification task of speech vs silence is stored in the temporary `silenceModelTemp.mat` file. The k-NN classifier is used to classify all the remaining mid-term segments (60%) of the input signal. For each mid-term segment, two estimates of posterior probabilities are generated. The value of k is selected to be equal to a predefined percentage of the number of samples of the training dataset.
5. At a post-processing stage, function `segmentationProbSeq()` is called in order to return the final segments. Both naive merging and Viterbi-based smoothing can be used at this stage.

The speech–silence segmenter has been implemented in the `silence Removal()` m-file. Figure 6.5 presents the results of the application of the method on a speech signal of short duration. For the sake of comparison, both post-segmentation approaches have been employed (naive merging and Viterbi-based smoothing).

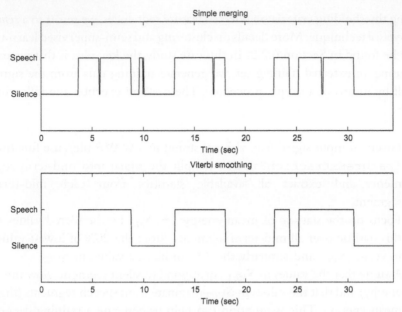

Figure 6.5 Speech-silence segmenter applied on a short-duration signal. The upper figure demonstrates the segmentation results after a naive merging post-processing stage has been performed. The lower figure presents the sequence of labels at the output of the Viterbi-based approach.

Note: Due to the fact that function `silenceRemoval()` generates a list of pairs of endpoints for the detected segments, along with the respective class labels (silence and speech), its results can be visualized using the previously described auxiliary plotting function `segmentationPlotResults()`. An example of use is given in the following piece of code:

```
% run the speech—silence segmentation method:
[segs, classes, Labels, centers] = ...
   silenceRemoval('example.wav', 0, 0);
% Plot (and possibly listen to) the results
segmentationPlotResults(segs, classes, 'example.wav');
```

6.1.1.4. Example: Fixed-Window Multi-Class Segmentation

This section presents a segmentation example that involves the classes of silence, male speech, female speech, and music. The training data that we need are stored in the `model4.mat` file. The segmenter employs a multi–class *k*-NN classifier on a mid-term segment basis. We have created (and stored in a WAVE file) a synthetic audio signal in which all the above classes are present. The ground-truth endpoints of the segments and the respective class labels are stored in a separate mat file for evaluation purposes. We use function `segmentationCompareResults()` to visualize the results and evaluate them against the ground truth. We also plot the posterior probability estimates for each class and each mid-term window. The complete example is implemented in the `demoSegmentation2.m` file and is presented below:

```
fileName = '../data/4ClassStream.wav';
gtFileName = '../data/4ClassStreamGT.mat'; kNN = 11;
% 4-class demo:
% Step 1: mid-term classification (classifier trained on 4 classes):
[labels, Ps, Conf, centers, classNames] = mtFileClassification(...
    fileName, kNN, 'model4.mat');
% Plot the class probabilities:
plot(centers, Ps); xlabel('Time (secs)'); title('Class Probabilities');
legend(classNames, 'Location', 'EastOutside','Interpreter', 'none');
load(gtFileName); % Load ground truth
% Step 2: extract segment boundaries:
[a, fs] = wavread(fileName, 'size'); totalDuration = a(1) / fs;
% 2A: naive merging:
[segs, classes, Labels] = segmentationProbSeq(...
    Ps, centers, totalDuration, 0); % segmentation (merging)
figure; segmentationCompareResults(segs, classes, ...
    segsReal, classesReal, fileName, {'Results','Ground-truth'});
subplot(2,1,1);
colorbar('Ytick',1:length(classNames), 'YTickLabel', classNames)
subplot(2,1,2);
colorbar('Ytick',1:length(classNames), 'YTickLabel', classNames)
% 2B: probability smoothing
[segs, classes, Labels] = segmentationProbSeq(...
    Ps, centers, totalDuration, 1); % segmentation (probability smoothing)
figure; segmentationCompareResults(segs, classes, ...
    segsReal, classesReal, fileName, {'Results','Ground-truth'});
subplot(2,1,1);
colorbar('Ytick',1:length(classNames), 'YTickLabel', classNames)
subplot(2,1,2);
colorbar('Ytick',1:length(classNames), 'YTickLabel', classNames)
```

The example calls the `segmentationProbSeq()` function twice, once for each post-processing method (naive merging and Viterbi-based smoothing). The results are illustrated in Figure 6.6. In this example, the Viterbi-based scheme outperforms the naive merger. More specifically, it can be observed that at the output of the Viterbi-based smoother only a small segment of male speech has been misclassified as silence, whereas, the naive merging scheme has misclassified 2 s of male speech as female speech.

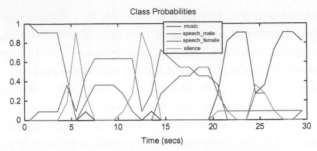

(a) Class probability estimates for each fixed window of the audio stream.

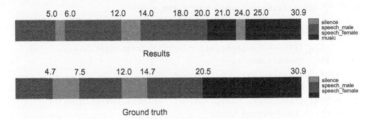

(b) Results of the naive merger in comparison with the ground truth.

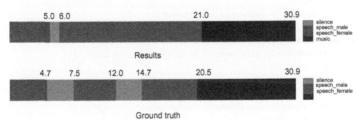

(c) Results of the Viterbi-based smoother in comparison with the ground truth.

Figure 6.6 Fixed-window segmentation with an embedded 4-class classifier (silence, male speech, female speech, and music).

6.1.2. Joint Segmentation-Classification Based on Cost Function Optimization

In this section, audio segmentation is treated as a maximization task, where the solution is obtained by means of dynamic programming [24]. The segmenter seeks the sequence of segments and respective class labels that *maximize the product of posterior class probabilities, given the data that form the segments.* Note that an increased understanding of this method can be gained if you consult, in parallel, Section 7.3 with respect to the Viterbi–related issues that are also involved in the current section.

To proceed further, some assumptions and definitions must be given. Specifically, it is assumed that each detected segment is at least T_{dmin} frames long, where T_{dmin} is a user-defined parameter. This is a reasonable assumption for the majority of audio types. Furthermore, it is assumed that the duration of a segment cannot exceed T_{dmax} frames. As will be made clear later in this section, this is a non-restrictive assumption dictated by the dynamic programming nature of the segmenter; any segment longer than T_{dmax} will be partitioned in segments shorter than T_{dmax} frames.

Let L be the length of a sequence, $\mathbf{X}$, of feature vectors, that has been extracted from an audio stream. If $\{d_1, d_2, \ldots, d_{K-1}, d_K\}$ are the frame indices that mark the boundaries of the segments, then a sequence of K segments can be represented as a sequence of pairs:

$$\{(1, d_1), (d_1 + 1, d_2), \cdots, (d_{K-1} + 1, L)\},$$

where $T_{dmin} \leq d_1 < d_2 \ldots < d_K = L$ and $T_{dmax} \geq d_k - d_{k-1} \geq T_{dmin}, k = 2, \ldots, K$. In addition, if C_k is the class label of the kth segment, then $p(C_k \mid \{X_{d_{k-1}+1}, \ldots, X_{d_k}\})$ is the posterior probability of the class label C_k given the feature sequence within the kth segment. For any given sequence of K segments and corresponding class labels, we can now form the cost function, $J(\cdot)$ as follows:

$$J(\{d_1, \ldots, d_K\}, \{C_1, \ldots, C_K\}, K) \equiv p(C_1 \mid \{X_1, \ldots, X_{d_1}\})$$
$$\times \prod_{k=2}^{K} p(C_k \mid \{X_{d_{k-1}+1}, \ldots, X_{d_k}\}) \qquad (6.1)$$

Function J needs to be maximized over all possible $K, \{d_1, d_2, \ldots, d_{K-1}, d_K\}$ and $\{C_1, C_2, \ldots, C_{K-1}, C_K\}$, under the two assumptions made in the beginning of this section. Therefore, the number of segments, K, is an outcome of the optimization process. Obviously, an exhaustive search is computationally prohibitive, so we resort to a dynamic programming solution. To simplify our presentation, we focus on a speech/music (S/M) scenario, although the method is valid for any number of classes.

We first construct a grid by placing the feature sequence on the x-axis and the classes on the y-axis (Figure 6.7). A node, (X_{d_k}, S), $T_{dmin} \leq d_k \leq L$, stands for the fact that a speech segment ends at frame index d_k (a similar interpretation holds for the (X_{d_k}, M) nodes). As a result, a path of K nodes, $\{(X_{d_1}, C_1), (X_{d_2}, C_2), \ldots, (X_{d_K}, C_K)\}$, corresponds to a possible sequence of segments, where $T_{dmin} \leq d_1 < d_2 < d_k = L$, $T_{dmax} \geq d_k - d_{k-1} \geq T_{dmin}, k = 2, \ldots, K$, and $\{C_1, \ldots, C_K\}$ are the respective class labels. The

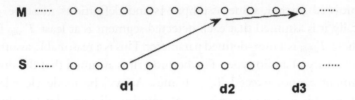

Figure 6.7 A sequence of segments in the dynamic programming grid.

transition $(X_{d_{k-1}}, C_{k-1}) \rightarrow (X_{d_k}, C_k)$ implies that a segment with class label C_{k-1} ends at frame d_{k-1}; and the next segment in the sequence starts at frame $d_{k-1} + 1$, ends at frame d_k and has class label C_k. We then associate a cost function, $T(\cdot)$, with the transition $(X_{d_{k-1}}, C_{k-1}) \rightarrow (X_{d_k}, C_k)$, as follows:

$$T((X_{d_{k-1}}, C_{k-1}) \rightarrow (X_{d_k}, C_k)) = p(C_k \mid \{X_{d_{k-1}+1}, \ldots, X_{d_k}\}) \qquad (6.2)$$

In other words, the cost of the transition is set equal to the posterior probability of the class label, C_k, given the feature sequence defining the segment $\{X_{d_{k-1}+1}, \ldots, X_{d_k}\}$.

Based on Eqs. (6.1) and (6.2), the cost function, $J(\cdot)$, becomes:

$$p(C_1 \mid \{X_1, \ldots, X_{d_1}\}) \cdot \prod_{k=2}^{K} T((X_{d_{k-1}}, C_{k-1}) \rightarrow (X_{d_k}, C_k))$$
$$= J(\{d_1, \ldots, d_K\}, \{C_1, \ldots, C_K\}, K) \qquad (6.3)$$

According to Eq. (6.3), the value of $J(\cdot)$ for a sequence of segments and corresponding class labels *can be equivalently computed as the cost of the respective path of nodes in the grid*. This means that the optimal segmentation can be treated as a best-path sequence on the grid. The rest of the solution is Viterbi-based processing, the principles of which are further explained in Section 7.3.

Note that the method requires estimates of the posterior probabilities, $p(C_k \mid \{X_{d_{k-1}+1}, \ldots, X_{d_k}\})$, of the class labels given the candidate segments. In [24], a Bayesian network combiner is used as a posterior probability estimator but this is not a restrictive choice. We can also use a k-NN scheme that operates on mid-term statistics drawn from the feature sequence of each segment, as has already been demonstrated in previous sections. At the end of this chapter you are asked to implement this dynamic programming-based segmenter as an exercise. You will need to consult Section 7.3 to gain an understanding of those dynamic programming concepts that underlie the operation of the Viterbi algorithm.

6.2. SEGMENTATION WITHOUT CLASSIFICATION

Section 6.1 presented segmentation techniques that rely on a trained classifier. However, this is not feasible in every application context. Consider, for example, the problem of speaker clustering (also known as speaker diarization, [64, 68]), where the goal is to partition a speech signal into homogeneous segments with respect to speaker identity. In such problems, there is no prior knowledge regarding the identity of the speakers, so it is not feasible to employ a classifier during the segmentation procedure.

In order to provide a solution, we resort to *unsupervised segmentation* approaches. The techniques that fall in this category can be further divided into two groups:

- *Signal change detection* methods: The output simply consists of the endpoints of the detected segments. Information related to the respective labels is not returned by the algorithm.
- *Segmentation with clustering*: The detected segments are clustered and the resulting clusters are used to assign labels. Speaker diarization solutions fall into this group of methods.

6.2.1. Signal Change Detection

The core idea of signal change detection is to determine when significant changes occur in the content of the audio signal. The detected changes define the boundaries of the resulting segments. Depending on the nature of the signals being studied (e.g. audio from movies, [69], music signals [70], etc.), the detection of endpoints may use a wide range of features and detection techniques.

As an example of this class of methods, we now present an outline of the most important algorithmic steps of an unsupervised detector of signal changes. It is assumed that the input to this method is just the audio signal and that the output consists of a list of pairs of endpoints.

1. Extract a sequence of mid-term feature vectors. This can be achieved with functions `stFeatureExtraction()` and `mtFeature Extraction()` which have already been described.
2. For each pair of successive feature vectors, compute a dissimilarity measure (distance function). As a result, a sequence of distance values is generated.
3. Detect the local maxima of the previous sequence. The locations of these maxima are the endpoints of the detected segments.

> *Note:* The above algorithm adopted the mid-term feature statistics because we were mostly interested in segmenting audio signals on a mid-term basis. If the nature of the application requires a more detailed segmentation scheme with respect to the time resolution of the detected endpoints, then, short-term feature vectors can be used directly in the computation of the distance metric. However, a large number of local maxima may be detected, leading to oversegmentation, unless a sophisticated peak selection technique is adopted.

The above distance-based approach to audio segmentation is implemented in the `segmentationSignalChange()` m-file. At the first stage, the function computes mid-term feature statistics for the audio content of a WAVE file:

```
[midFeatures, Centers, stFeaturesPerSegment] = ...
    featureExtractionFile(fileName, stWin, stStep, mtWin, mtStep, featureStatistics);
```

The extracted features are then normalized to zero mean and standard deviation equal to one. This linear normalization step prevents features from dominating the computation of the distance measure, i.e. the resulting values are not biased toward features with large values.

```
midFeatures = midFeatures';
MEAN = mean(midFeatures); STD = std(midFeatures);
midFeatures = (midFeatures - repmat(MEAN, [size(midFeatures,1) 1])) ...
    ./ repmat(STD, [size(midFeatures,1) 1]  );
```

The distance metric is computed using the `pdist2()` MATLAB function, for each row, `i`, of the feature matrix:

```
Dist(i) = (pdist2(midFeatures(i, :), midFeatures(i-1, :)));
```

Then, we use the `findpeaks()` MATLAB function of the Signal Processing Toolbox to detect the local maxima of the sequence of distance values:

```
[maxDist, iMaxDist] = findpeaks(Dist);
```

As a final step, the peaks whose values are higher than a threshold are preserved. The value of the threshold is equal to the mean value of the sequence of dissimilarities. The use of a threshold filters out local maxima with low values.

We now provide an example of the execution of the `segmentation-SignalChange()` algorithm on an audio signal from a TV program. In order to demonstrate the performance of the function, we have manually annotated the boundaries of segments of the input stream. In addition, we have used function `segmentationCompareResults()` to visualize the extracted segments and compare them to the results of the manual annotation procedure. To reproduce the example, type the code that is given below (or execute the `demoSegmentationSignalChange` m-file). The code first loads a `.mat` file with the true segment boundaries and respective class labels (the latter only serving as auxiliary information in this case).

```
% load the groundtruth:
load ../data/topGearGT.mat
% signal change detection:
[segs] = segmentationSignalChange('../data/topGear.wav', 1);
% class labels are not available here, so just
% use labels of alternating values (0, 1, 0, 1, etc):
for i=1:size(segs,1) classes(i) = mod(i,2); end
% plot results
titles{1} = 'Estimated'; titles{2} = 'Ground truth';
segmentationCompareResults(segs, classes, ...
    segsReal, classesReal, '../data/topGear.wav', titles);
```

Figure 6.8 presents the results of the segmenter against the manually annotated stream. It can be seen that in most cases the extracted segment limits are in close agreement with the ground truth (label information is ignored in this task, since we are only interested in the segments' limits and not in their labels).

6.2.2. Segmentation with Clustering

The demand for unsupervised segmentation is also raised by certain audio analysis applications that additionally require that the segments are clustered to form groups, so that the segments of each cluster (group) are automatically assigned the same label. Speaker diarization falls in this cat-

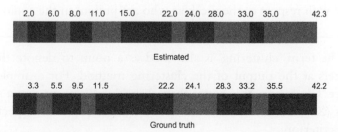

Figure 6.8 Top: Signal change detection results from a TV program. The colors are for illustration purposes and do not correspond to class labels. B: Ground truth. The colors represent different classes but this type of information can be ignored during the evaluation of the performance of the segmenter.

egory of applications; each cluster refers to a different speaker. The output of a diarization system can be seen as a sequence of segments, where each segment has been automatically annotated with the identity (label) of the cluster to which it belongs, e.g., $((t_1, t_2), \text{speaker A}), ((t_3, t_4), \text{speaker B}), ((t_5, t_6),$ speaker A$), ((t_7, t_8), \text{speaker C})$, where the t_i s refer to segment boundaries. In this example, three clusters are formed and one of them, tagged (speaker A), consists of two segments, namely $((t_1, t_2), \text{speaker A})$ and $((t_5, t_6), \text{speaker A})$. In order to understand how speaker diarization and related methods operate, it is firstly important to become familiar with data clustering and related concepts.

6.2.2.1. A Few Words on Data Clustering

So far, we have mainly focused on supervised tasks, where each sample was assigned to a set of predefined audio classes. We are now facing the challenge of dealing with unlabeled data. Our goal is to determine, automatically, the structure of the data. In other words, we are interested in partitioning the dataset of unlabeled samples to a number of subsets, so that the data in each subset are similar to one another and at the same time sufficiently different from the data in other subsets. The procedure that extracts this type of knowledge from a dataset is known as *clustering* and the resulting subsets as *clusters* [5, 38, 71–73]. Clustering algorithms require a definition of similarity, so that the generated clusters 'make sense.' Due to the fact that similarity can be defined at different levels of abstraction and can address different types of data, it is important to understand that, depending on the application, the adopted similarity measure must serve our needs to optimize the clustering procedure with respect to some appropriately defined criteria. For example, in the case of speaker diarization, a similarity measure needs to be defined among speakers, and the goal of the clustering scheme is to produce a set of clusters that is optimal with respect to the speakers who are present in the audio stream.

Note: The term 'clustering' is also used as a noun to denote the set of clusters at the output of the clustering method. For example, the sentence 'Clustering X is better than clustering Y' means that a set of clusters (X) is better than another set of clusters (Y) with respect to a quality criterion.

A clustering example is shown in Figure 6.9, for an artificially generated dataset consisting of data from four two-dimensional Gaussian distributions.

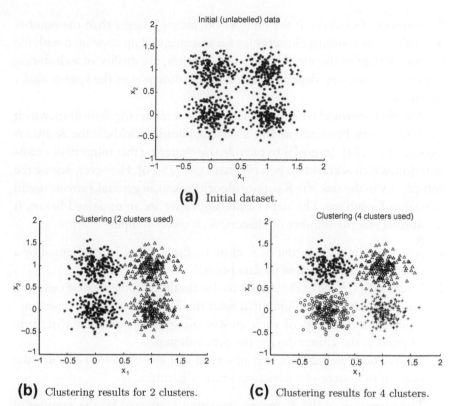

(a) Initial dataset.

(b) Clustering results for 2 clusters.

(c) Clustering results for 4 clusters.

Figure 6.9 A clustering example in the two-dimensional feature space. The two results correspond to different numbers of clusters (2 and 4). The figures have been generated using the provided demoCluster_general() function.

The clustering algorithm was run twice and each time the desired number of clusters was given as input, as it was dictated by the nature of the clustering algorithm that we used (the k-means algorithm). Figures 6.9b and 6.9c present the clustering results for two and four clusters, respectively. Note that, from a visualization perspective, the second clustering seems to provide a better description of the dataset, because the data from each distribution form a separate cluster. However, in practice, it is not always obvious which clustering is a better representative of the dataset and in most cases it is an application-related issue. In a speaker diarization scenario, several different clusterings can be considered to be reasonable depending on the requirements. For example, it makes sense to ask the algorithm to generate two clusters, if we want to split the recording based on the gender of the speaker. A larger number of clusters can be given as input if we are interested in detecting who spoke when and we have obtained prior knowledge about the number of speakers. If the number of clusters is larger than the number of speakers, the resulting clusters may, for instance, end up associated with the emotional state of the speaker(s). In other words, the quality of a clustering can be subjective and depends on the desired outcome of the system under development.

A well-known and computationally efficient clustering algorithm, which has traditionally been adopted in diverse application fields, is the K-means algorithm [71, 74]. Its goal is to provide the clustering that minimizes a cost-function, which is known to be a NP-hard task [75, 76]. However, despite the complexity of the task, the K-means algorithm can, in general, provide useful suboptimal solutions. The steps of the algorithm are summarized below. It is assumed that the number of clusters, K, is given as input:

1. Select an initial clustering of K clusters. Each cluster is represented by a *mean vector*, the *centroid* of its data points.
2. Assign each sample of the dataset to the cluster with the nearest centroid. The proximity can be computed with the Euclidean distance function.
3. Update the centroid of each cluster using all the samples that were assigned to the cluster during the previous step.
4. Repeat from step 2 until it is observed that the clustering remains the same in two successive iterations of the algorithm.

The operation of the K-means algorithm is affected by the number of clusters at the input, the initialization scheme (initial clustering), and the choice of distance metric. The estimation of the number of classes is not a trivial issue. A possibility is to run the K-means algorithm for a range of values of the K parameter and select the 'best' solution after evaluating each clustering against a quality criterion [77, 78]. The task of estimating the number of clusters is presented in more detail in Section 6.2.2.1.1. The initialization of the algorithm is also of crucial importance for its convergence to an acceptable solution. A well-known technique is to initialize the algorithm randomly a number of times and select as the winner the most frequently appearing clustering or the clustering that corresponds to the lowest value of the cost function.

The K-means algorithm is a typical centroid-based technique because each cluster is represented by its centroid (note that centroids do not necessarily coincide with points of the dataset). Another important category of clustering methods consists of the so-called *hierarchical clustering* algorithms. The goal of these algorithms is to produce a hierarchy of clusters given the points of a dataset. The extracted hierarchy is presented as a tree-like structure [72]. The two dominant trends in the field of hierarchical clustering are the *agglomerative* and *divisive* techniques:

- *Agglomerative methods.* In the beginning of the algorithm, the number of clusters is equal to the number of samples, i.e. each cluster contains only one sample. At each iteration, the two most similar clusters are merged and the procedure is repeated until a termination criterion is satisfied [79]. The agglomerative algorithms follow a bottom–up approach because the number of clusters is reduced by one at the end of each iteration.
- *Divisive methods.* These algorithms adopt a top-down approach. They start with a root cluster that contains the entire dataset and perform an iterative splitting operation based on an adopted criterion [80, 81].

In recent years, we have also witnessed the increasing popularity of *spectral* clustering techniques [73, 82–84]. The key idea behind spectral clustering is that the dataset is first represented as a weighted graph, where the vertices correspond to data points and the edges capture similarities between data points. Clustering is then achieved by cutting the graph in to disconnected components so that an optimization criterion is fulfilled. Another modern approach to data clustering is provided by the Non–Negative Matrix Factorization (NMF) methodology, which approaches the problem of clustering from the perspective of *concept* factorization [73, 85–88]. NMF techniques treat each data point as a linear combination of all cluster centers. Applications of NMF include sound source separation [89] and document clustering [85].

6.2.2.1.1. Estimating the Number of Clusters

Several solutions have been proposed in the literature for the estimation of the number of clusters in a dataset [77, 90–92]. A popular approach is to execute the K-means algorithm for a range of values of parameter K and select the optimal clustering, based on a selected criterion, like the 'gap statistic' [77]. In this section, we have adopted the *Silhouette* method proposed in [92]. This is a simple method that measures how 'tightly' the data are grouped in each cluster, using a distance metric. The Silhouette method is implemented in the `silhouette()` function of the Statistics Toolbox. Figure 6.10 presents a clustering example where the Silhouette method was used to estimate the number of clusters. It can be observed that the Silhouette measure is maximized when the number of clusters is 4 and this is the adopted solution in this case. Note that some clustering approaches, e.g. the hierarchical clustering methods, may embed the estimation of the number of clusters in the clustering procedure itself.

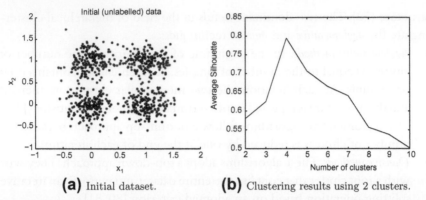

(a) Initial dataset.   **(b)** Clustering results using 2 clusters.

Figure 6.10 Silhouette example: the average Silhouette measure is maximized when the number of clusters is 4. This example was generated using the `demoCluster_generalSil()` function that can be found in the software library that accompanies this book.

6.2.2.1.2. Data Clustering Functionality in MATLAB

MATLAB includes several functions that perform data clustering. In this section we provide a brief description of the K-Means algorithm and the agglomerative clustering technique.

The K-Means algorithm has been implemented in the `kmeans()` function of the Statistics Toolbox. The function takes as input a data matrix with one feature vector per row along with the desired number of clusters. It returns a vector of cluster labels for the data points, as well as the centroids of the clusters in a separate matrix. If the user demands a deeper understanding of the operation of the algorithm, more output arguments are possible, e.g. the within-cluster sums of point-to-centroid distances.

The Statistics Toolbox also supports the agglomerative (hierarchical) clustering method. To use the respective functionality, take the following steps:

1. Use the `pdist()` MATLAB function to compute the similarity (or dissimilarity) between every pair of samples of the dataset.
2. Form a hierarchical tree structure using the `linkage()` function, which provides the agglomerative functionality. The resulting structure is encoded in a specialized matrix format.
3. Use the `cluster()` function to transform the output of the `linkage()` function to a set of clusters based on a thresholding criterion.

The above three-step agglomerative procedure can also be executed in a single step using the MATLAB function `clusterdata()`. We have provided a stepwise description so that the reader becomes familiar with the key mechanism on which agglomerative clustering operates.

6.2.2.1.3. Semi-Supervised Clustering

In many clustering applications, although the dataset is unlabeled, there exists auxiliary information that can be exploited by the clustering algorithm to yield more meaningful results. Such methods are usually referred to as *semi-supervised* clustering techniques [71, 93, 94].

A popular approach to semi-supervised clustering is to make use of *pairwise constraints*. For example, a 'must-link' constraint can specify that two samples from the dataset must belong to the same cluster. This type of constraint has been used in the field of speaker clustering. The work in [64] proposes a semi-supervised approach that groups speech segments to clusters (speakers) by making use of the fact that two successive segments have a high probability of belonging to the same speaker. This auxiliary information has been proven to boost the performance of the clustering scheme.

> *Note:* Semi-supervised clustering should not be confused with semi-supervised classification. The latter refers to the situation where some of the data are labeled and during the training stage the unlabeled data are also used by the learning algorithm

6.2.2.2. A Speaker Diarization Example

In this section, we provide the description and respective MATLAB code of a simple speaker diarization system. Remember that speaker diarization is the task of automatically partitioning an audio stream into homogeneous segments with respect to speaker identity [64, 68, 95]. The input of a speaker diarization (SD) method is a speech signal, without any accompanying supervised knowledge concerning the speakers who are involved. The goal of a SD system is to:

- Estimate the number of speakers.
- Estimate the boundaries of speaker-homogeneous segments.
- Map each segment to a speaker ID.

For the sake of simplicity, we will only describe the main speaker diarization functionality, assuming that *the number of speakers is provided by the user*. Of course, generally in the general case, the number of speakers is unknown and has to be estimated using a suitable method like the one in Section 6.2.2.1.1. A schematic description of our diarization method is shown in Figure 6.11. The respective source code can be found in the `speakerDiarization()` function of the software library that accompanies this book. The method involves the following steps:

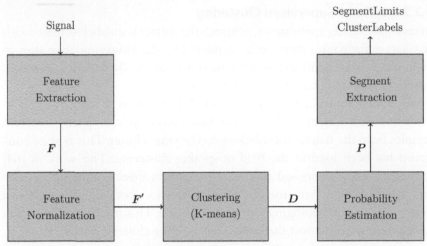

Figure 6.11 Block diagram of the speaker diarization method implemented in speak-erDiarization(). **F** is the matrix of feature statistics of all segments and **F'** is its normalized version. **D** is the distance matrix, which is returned by the **K**-means algorithm. **P** is the matrix of probability estimates, which is used in the final segmentation stage to generate the detected segment boundaries and corresponding cluster labels.

1. Computes feature statistics on a fixed-length segment basis using the featureExtractionFile() function already used several times in this book. The result of this procedure is a sequence of vectors of feature statistics and a sequence of time stamps (in seconds), which refer to the time equivalent of the centers of the mid-term segments.
2. The feature statistics, which were extracted in the first step, are linearly normalized to zero mean and unity standard deviation. The normalization step increases the robustness of the clustering procedure.
3. The (normalized) feature statistics are used by the K-Means clustering algorithm, which assigns a cluster identity to each mid-term segment. Apart from the segment labels, the clustering algorithm also returns the distances of the feature vectors (representing the segments) from all the centroids. This is the 4th output argument of the kmeans() function of the Statistics Toolbox. The computed distances form a $[C \times T]$ matrix, D, where C is the number of clusters and T is the number of segments.
4. The distance matrix is used to provide a crude approximation of the probability that a segment belongs to a cluster. This is achieved by computing the quantity $1/D(i, j)$ for each element of D and then by normalizing the resulting values column-wise. The result of this step is a $[C \times T]$ matrix P, whose elements, $P(i, j)$, represent the estimated probability that sample

(segment) j belongs to cluster i. Each row of this matrix is subsequently smoothed using a simple moving average technique.

5. The probabilistic estimates are then given as input to the `segmentationProbSeq()` MATLAB function (already described in Section 6.1.1.1), which produces the final list of segments and respective cluster labels.

The following code demonstrates how to call the `speakerDiarization()` function and plot its results using the `segmentationPlotResults()` function:

```
% Run the speaker diarization function
% Prior knowledge is provided regarding the number of speakers (4):
[segs, classes] = speakerDiarization('../data/diarizationExample', 4);
% Plot (and listen to) the results
segmentationPlotResults(segs, classes, '../data/diarizationExample);
```

Figure 6.12 presents the results of the application of the speaker diarization method. Note that the `speakerDiarization()` function implements a typical clustering-based audio segmentation technique. The reason that we have focused on speaker diarization is that it is a well-defined task with meaningful results. The same function can be used in any clustering-based audio segmentation application. The reader is encouraged to test the function on other types of audio signals, e.g. on a music track, and provide an interpretation of the resulting segments. In general, music

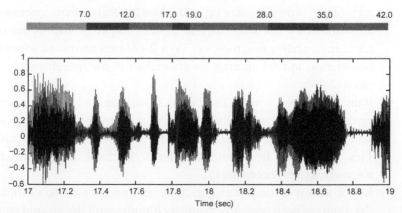

Figure 6.12 Visualization of the speaker diarization results obtained by the `speakerDiarization()` function (visualization is obtained by calling the `segmentationPlotResults()` function). Different colors represent different speaker labels. The lower part of the figure represents the waveform of an audio segment, which was selected by a user.

segmentation is a very different task from speaker diarization, requiring different features and techniques. More on selected music analysis applications are presented in Chapter 8.

6.3. EXERCISES

1. (D3) Follow the instructions given below in order to create a MATLAB function that implements a speech-silence segmenter that employs a binary classifier:
   - Load the model4.mat file, keep the data that refer to the classes of speech and silence, and discard the rest. Create a classification setup for the speech vs silence task and store the data in the modelSpeechVsSilence.mat file.
   - Reuse the code that was presented in this chapter to design a fixed-window segmenter that adopts a k-NN classifier.
   - Record a speech signal to demonstrate your implementation and compare the results with those obtained by the silenceRemoval() function of Section 6.1.1.3.

2. (D3) Section 6.1.1.1 presented a segmenter that was based on a kNN classifier of fixed-length windows, using the speech-music dataset of the modelSM.mat file. In this exercise you are asked to:
   - Annotate manually the speech_music_sample.wav file from the data folder of the companion material. You will need to create a mat-file (speech_music_sampleGT.mat) that contains annotations of the same format as the output of the segmentationProbSeq function, i.e.: (a) a 2-column matrix of segment boundaries, and (b) an array of class labels (0 for speech and 1 for music).
   - Run functions mtFileClassification() and segmentationProbSeq() to segment the speech_music_sample.wav file using the modelSM.mat file for the classification task. This needs to be done for both methods supported by segmentationProbSeq().
   - Use function segmentationCompareResults() to compare the results of both post-processing techniques with the ground truth that you have created.

   Hint: In Section 6.1.1.4 we demonstrated the same task for a 4-class scenario. Here, you only need to create ground truth data for a WAVE file and then go through the steps of Section 6.1.1.4, using the modelSM.mat file.

3. (D5) Implement a variant of a region growing technique that splits an audio stream into speech and music parts [24]. The main idea is that each segment can be the result of a segment (region) growing technique, where one starts from small regions and keeps expanding them as long as a predefined threshold-based criterion is fulfilled. To this end, implement the following steps:

    (a) Extract the short-term feature of spectral entropy from the audio recording (see Chapter 4). Alternatively, you can use the entropy of the chroma vector.

    (b) Select a number of frames as candidates (seeds) for region expansion. The candidates should preferably be equidistant, e.g. one candidate every 2 s (user-defined parameter).

    (c) Starting from the *seeds*, segments grow to both directions and keep expanding as long as the standard deviation of the feature values in each region remains below a predefined threshold. Note that a feature value can only belong to a single region. Therefore, a region ceases to grow if both its neighboring feature values already belong to other regions. At this end of this processing stage, the resulting segments contain *music* and all the rest is *speech*.

    (d) Post-processing: adjacent segments are merged and short (isolated) segments are ignored. All segments that have survived are labeled as music, because the algorithmic parameters are tuned toward the music part of the signal.

Your function must receive as input a WAVE file, the type of feature, the distance between successive seeds (in seconds), and the threshold value for the region growing step. The segmentation results must conform with the output format that has been adopted in the chapter. Experiment with different threshold values and comment on the results. Compare the performance of your implementation with a fixed-window segmenter that uses a binary classifier for the task of speech vs music.

4. (D5) Implement the Viteri-based solution of Eq. (6.3). Use a k-NN estimator for the posterior probabilities. Your function must take as input the path to a WAVE file, the short-term features to be extracted, the segment duration constraints, and the mid-term statistics that will be used during the estimation of the posterior probabilities. You will need to consult Chapter 7 in order to gain an understanding of the dynamic programming concepts that underlie the operation of the Viterbi algorithm. On a 4-class problem, compare the results of this method with the performance of a fixed-window technique.

5. (D1) The core technique in the `speakerDiarization()` function is a generic clustering-based audio segmenter. Therefore, it makes sense to try and use it with non-speech signals. Experiment with the `speakerDiarization()` and `segmentationPlotResults()` functions on music tracks and assign meaning (if possible) to the clustering results.

6. (D3) Create a Matlab function that extends the functionality of the `speakerDiarization()` method by computing a *confidence measure* for each detected segment. Such a measure can be quite useful in various applications. For instance, one could use it to filter out segments that correspond to low confidence values, so that the algorithm outputs only those segments for which confidence is sufficiently high. Furthermore, the confidence measure can be adopted by fusion techniques (e.g. jointly with visual-oriented cues). *Hint: Compute the average estimated probability (using the `Ps` matrix in the `speakerDiarization()` function) at each segment over all fixed-sized windows contained in that segment.*

7. (D5) Assume that, in a speaker diarization context, the number of speakers has not been provided by the user. Extend the functionality of the `speakerDiarization()` function by adding an estimator for the number of clusters (speakers). To achieve this goal, use the Silhouette method that was explained in Section 6.2.2.1.1. The new function must receive a single input argument, i.e. the path to the audio file to analyze.

8. (D4) The `speakerDiarization()` function estimates, for each segment, the probability that the segment belongs to each cluster. However, this is achieved in a rather simple manner (using the inverse distance from each cluster centroid). Modify the `speakerDiarization()` function so that the probability estimate is computed by means of a more sophisticated technique. This can be obviously achieved in many ways and the reader can choose any valid probability estimation method. Here, we propose that you use a k-NN classifier that exploits the centroids of the clusters at the output of the K-means algorithm in order to estimate the probability that each segment belongs to a cluster.

9. (D4) The `speakerDiarization()` function assumes that the input signal contains only speech. However, if long intervals of silence/background noise are also present, reduced performance may be observed. Combine `speakerDiarization()` and `silenceRemoval()` to filter out areas of silence from the input signal before the clustering scheme is employed.

10. (D3) The signal change detection approach in Section 6.2.1 extracts segments of homogeneous content without making use of any type of

supervised knowledge. In several audio applications, such a functionality can be used as a first step in a segmentation – classification system. In this exercise, combine:

- Function `segmentationSignalChange()` to extract homogeneous audio segments from an audio stream.
- The feature extraction and classification process (see Chapter 5) to classify (k-NN) each extracted segment as music or speech (use `modelSM.mat`).

Hint: The feature extraction and classification processes must be applied on each segment (extracted by the signal change detection step), using the following sequence of operations: (a) short-term feature extraction, (b) mid-term statistics generation, (c) long-term averaging of the statistics for the whole segment, and (d) application of the classifier. The same sequence of steps was followed in the `fileClassification()` *function.*

CHAPTER 7

Audio Alignment and Temporal Modeling

Contents

This chapter presents several methods that take into account the temporal evolution of the audio phenomena. In other words, we are no longer interested in computing mid-term feature statistics or long-term averages from the audio signals. On the contrary, our main concern is to preserve the short-term nature of the feature sequences in order to (a) devise techniques that are capable of aligning two feature sequences, and (b) build parametric, temporal representations of the audio signals by means of the Hidden Markov modeling methodology. The two goals are complementary in the sense that the former involves non-parametric approaches, whereas the latter aims at deriving parametric stochastic representations of the time-varying nature of the signals. The choice of method depends on the application and the availability of training data. Sometimes non-parametric techniques are the only option, as is, for example, the case with query-by-example scenarios where the user provides an audio example that needs to be detected in a corpus of audio recordings. On the other hand, if sufficient training data are available, it might be preferable to build parametric models, as has traditionally been the case in the speech recognition literature. The chapter starts with sequence alignment techniques and proceeds with a description of the Hidden Markov models (HMMs) and related concepts. We have found it necessary to provide lengthier theoretic descriptions in this chapter. However, care has been taken, so that the presentation is given from an engineering perspective, in order to facilitate the realization of practical solutions.

7.1. AUDIO SEQUENCE ALIGNMENT

In the field of content-based audio retrieval, it is often useful to detect occurrences of an audio pattern in an audio stream, match audio data of varying lengths to a set of predefined templates, compare audio recordings to identify similar parts, and so on. The related terminology has its origins in speech processing literature, where similar problems have been intensively studied over past decades. For example, in an analogy of the *word spotting problem* in the speech recognition context, we can define the *audio spotting* or *audio tracking* task, where the goal is to detect occurrences of an audio example in a corpus of audio data. The audio example can be a digital audio effect that we need to track in the audio stream of a movie or a specific sound of a mammal that is likely to be repeated in a corpus of underwater recordings of very long duration. An endless list of similar examples can be compiled for a diversity of applications, ranging from intrusion detection systems to query-by-humming services. It is important to realize that we are interested in providing non-parametric solutions, i.e. we assume that we have not previously built parametric models that describe the problem at hand, based on prior knowledge.

An important requirement in such application scenarios is the ability to *align* feature sequences. Although in the literature the term *sequence alignment* refers to a broad class of problems, we start our discussion with the well-known *template matching* task, a direct application of which, can be encountered in the design of classifiers, when each class can be represented by one or more *reference patterns* (*prototypes*). The key idea behind template matching is that we can quantify how similar (dissimilar) two feature sequences are, even when they are not of equal length. For example, if you record your own voice repeating the word 'library' ten times with sufficiently long periods of silence between successive utterances and examine the resulting signals, you will observe that no two words are alike at the signal level. Some phonemes will be of shorter duration, whereas others will last longer, even if the differences are in the order of a few milliseconds. Furthermore, the same phoneme can be louder in some utterances. Yet, all signals are easily perceived by a human as the word 'library' and, in addition, if a different word is uttered, we will immediately percieve the difference.

7.1.1. Dynamic Time Warping

In order to deal with the alignment task from an algorithmic perspective, we understand that we need algorithms that are capable of quantifying similarity (dissimilarity) in an optimal way, by taking into account the fact that parts of one sequence may need to be locally stretched or compressed, so that the optimal alignment is achieved. The operations of stretching and compressing audio parts are collectively known as *time warping*. Due to the fact that time warping is determined dynamically during the joint inspection of two feature sequences, the term *Dynamic Time Warping (DTW)* is often used to denote the alignment of two feature sequences that represent audio signals.

To proceed further, certain definitions must first be given. Let $\mathbf{X} = \{\underline{x}(i); i = 1, \ldots, T_x\}$ and $\{\mathbf{R} = \underline{r}(j); j = 1, \ldots, T_r\}$, be the two feature sequences to be aligned. In general, $T_x \neq T_r$ and $\underline{x}(i)$ and $\underline{r}(j)$ are K-dimensional feature vectors. Furthermore, let $d(\underline{a}, \underline{b})$ be a dissimilarity function, e.g. the Euclidean distance function, which is defined as

$$d(\underline{a}, \underline{b}) = \sqrt{\left(\sum_{l=1}^{K} (\underline{a}_l - \underline{b}_l)^2 \right)}. \tag{7.1}$$

The DTW algorithms start with the construction of a *cost grid*, C. Assuming that sequences $\mathbf{R}$ and $\mathbf{X}$ are placed on the horizontal and vertical axis, respectively, the coordinates of node (i, j) are the ith element (feature vector) of $\mathbf{X}$ and the jth element of $\mathbf{R}$. Subsequently, a dynamic programming algorithm is employed. The algorithm starts from node $(1, 1)$ and the nodes of the grid are visited row-wise from left to right (an equivalent column-wise visiting scheme can also be adopted). At each node, we compute the accumulated cost to reach the node. At the end of this processing stage, the cost at node (T_x, T_r) represents the total matching cost between the two feature sequences. A zero cost indicates a perfect match. The higher the matching cost, the more dissimilar (less similar) the two sequences.

Obviously, there are several open questions: which are the nodes from which a node, (i, j), can be reached? What is the cost of a transition from one node to another? Why is this procedure capable of generating the optimal matching cost? To answer such questions, we now provide a step-by-step description of two popular DTW schemes, which are based on the so-called *Sakoe-Chiba* [96] and *Itakura* [97] local path constraints.

7.1.1.1. The Sakoe-Chiba Local Path Constraints

Initialization: We start with node $(1,1)$ and set $C(1, 1) = d(\underline{x}(1), \underline{r}(1))$. Due to the fact that node $(1,1)$ is the first node that we visit, we define its predecessor as the fictitious node $(0,0)$. We are using a separate $(T_x \times T_r)$ matrix, P_B, to store the (best) predecessor of each node. In this case, $P_B(1, 1) = (0, 0)$.

Iteration: We adopt a row-wise visiting scheme (from left to right at each row) and compute the cost, $C(i, j)$, to reach node (i, j), as follows:

$$C(i,j) = \min\{C(i-1, j-1), \ C(i-1, j), \ C(i, j-1)\} + d(\underline{x}(i), \underline{r}(j)). \quad (7.2)$$

Equation (7.2) defines the local path constraints of the DTW scheme. Specifically, it defines that the allowable predecessors of node (i, j) are the nodes: $(i - 1, j - 1)$ (diagonal transition); $(i - 1, j)$ (vertical transition); and $(i, j - 1)$ (horizontal transition). In addition, it defines that the cost of the transition from a predecessor to node (i, j) is equal to the local cost (dissimilarity) between $\underline{x}(i)$ and $\underline{r}(j)$. The node that minimizes Eq. (7.2) is the best predecessor of node (i, j) and its coordinates are stored in the P_B matrix. The value $C(i, j)$ stands for the optimal (lowest) cost to reach node (i, j), i.e. the more efficient way to reach the node, given the local path constraints. This is guaranteed by Bellman's optimality principle [98], which states that the optimal path to reach node (i, j) from $(1,1)$ is also optimal for each one of the nodes that constitute the path.

Termination: After the accumulated cost has been computed for all nodes, we focus on node (T_x, T_r). The cost to reach this node is the overall matching cost between the two feature sequences. To extract the respective optimal path, we perform a backtracking procedure: starting from node (T_x, T_r), we visit the chain of predecessors until we reach the fictitious node $(0,0)$, which is the predecessor of node $(1,1)$. The extracted chain of nodes is known as the *best (optimal) path* of nodes.

Horizontal and vertical transitions in the best path imply that certain time warping was necessary in order to align the two sequences; the longer the horizontal and vertical transitions, the more intense the time warping phenomenon. The Sakoe-Chiba constraints permit arbitrarily long sequences of horizontal or vertical transitions to take place, which implies that they do not impose any time warping limits. This is not always a desirable behavior, hence the need for other types of local constraints, like the Itakura constraints, which are described next.

We provide an implementation of the Sakoe-Chiba DTW scheme in the dynamicTimeWarpingSakoeChiba() function. To call this function, type:

```
[DTWCost,BestPath]=dynamicTimeWarpingSakoeChiba(B,A,tonorm);
```

The first two input arguments are the K-dimensional feature sequences to be aligned. Sequence B is placed on the vertical axis of the grid. The third argument is set to '1' if we want to divide the resulting matching cost (first output argument) by the length (number of nodes) of the best path. This type of normalization is desirable when the DTW scheme is embedded in a classifier that matches unknown patterns against a set of prototypes, because the normalized cost is an arithmetic average that is independent of the length of the sequences that are being matched. The second output argument is the sequence of nodes (pairs of indices) of the optimal alignment path.

7.1.1.2. The Itakura Local Path Constraints

In several audio alignment tasks it is undesirable to permit arbitrarily long sequences of horizontal and vertical transitions, because we expect that two instances of an audio pattern cannot be radically different. It is possible to account for this requirement, if we replace the Sakoe-Chiba scheme with a different set of local constraints. A popular choice are the *Itakura* local path constraints [97], which are defined as:

$$C(i,j) = \min\{C(i-1,j-1), C(i-2,j-1), C(i,j-1)\} + d(\underline{x}(i), \underline{r}(j)),$$
$$\text{if } P_B(i,j-1) \neq (i,j-2),$$
$$\text{and}$$
$$C(i,j) = \min\{C(i-1,j-1), C(i-2,j-1)\} + d(\underline{x}(i), \underline{r}(j)),$$
$$\text{if } P_B(i,j-1) = (i,j-2). \tag{7.3}$$

First of all, it can be easily observed that vertical transitions are not allowed at all. Secondly, it is possible to omit feature vectors from the sequence that has been placed on the vertical axis (one feature vector at a time), due to the fact that node $(i-2, j-1)$ is an allowable predecessor. A third important remark is that two successive horizontal transitions are not allowed due to the constraint that demands $P_B(i, j-1) \neq (i, j-2)$. These properties ensure that time warping cannot be an arbitrarily intense phenomenon. An implication of the Itakura constraints is that $\frac{1}{2}T_r \leq T_x \leq 2T_r$. Furthermore, the Itakura constraints are *asymmetric* because, in the general case, if the sequences change axes, the resulting matching cost will be different (on the contrary, the Sakoe-Chiba constraints are *symmetric*).

The Itakura algorithm is initialized like the Sakoe-Chiba scheme and Eq. (7.3) is repeated for every node. In the end, node (T_x, T_r) has accumulated the matching cost of the two sequences. The backtracking stage extracts the optimal sequence of nodes, assuming again that the best predecessor of each node has been stored in a separate matrix.

The software library that accompanies this book contains an implementation of the Itakura algorithm in the `dynamicTimeWarpingItakura()` function. To call this function, type:

```
[DTWCost,BestPath]=dynamicTimeWarpingItakura(B,A,tonorm);
```

For the sake of uniformity, we have preserved the input and output formats of the `dynamicTimeWarpingSakoeChiba()` function. Both functions have adopted the Euclidean distance as a dissimilarity measure.

7.1.1.3. The Smith-Waterman Algorithm

The Smith-Waterman (S-W) algorithm was originally introduced in the context of molecular sequence analysis [99]. It is a sequence alignment algorithm that can be used to discover similar patterns of data in sequences consisting of symbols drawn from a discrete alphabet.

An important characteristic of the S-W algorithm is that it employs a similarity function that generates a positive value to denote similarity between two symbols and a negative value for the case of dissimilarity. In addition, it introduces a gap penalty to reduce the accumulated similarity score of two symbol sequences when a part of one sequence is missing from the other. The algorithm is based on a standard dynamic programming technique to reveal the two subsequences that correspond to the optimal alignment. In the field of music information retrieval, the S-W algorithm has been used to provide working solutions for the tasks of cover song identification [100], query-by-humming [101], and melodic similarity [102]. In the more general context of audio alignment, variants of the algorithm that operate on a continuous feature space have been presented in [103,104].

In order to proceed with the description of the original S-W algorithm, let A and B be two symbol sequences, for which $A(i)$, $i = 1, \ldots, M$ and $B(j)$, $j = 1, \ldots N$, are the ith and jth symbol, respectively. We first need to define a function, $S(i, j)$, that takes as input two discrete symbols, $A(i)$, $B(j)$ and returns a measure of their similarity. The similarity function, $S(i, j)$, is defined as

$$S(i, j) = \begin{cases} +1 \text{ if } A(i) = B(j), \\ -\frac{1}{3} \text{ if } A(i) \neq B(j). \end{cases} \qquad (7.4)$$

At the next stage, we construct a $(M + 1) \times (N + 1)$ similarity grid, H. We assume that the rows and columns of the grid are indexed $0, \ldots, M$ and $0, \ldots, N$, respectively. Each node of the grid is represented as a pair of coordinates, (i, j), $i = 0, \ldots, M$, $j = 0, \ldots, N$. The basic idea of the algorithm is that each node accumulates a similarity score by examining a set of nodes known as *predecessors*. After the accumulated score has been computed for all the nodes of the grid, the node with the highest score wins and a *backtracking* procedure reveals the two symbol subsequences that have contributed to this maximum score.

Initialization: The first row and column of the grid are initialized with zeros:

$$H(0, j) = 0, j = 0, \ldots, N \quad and \quad H(i, 0) = 0, i = 0, \ldots, M.$$

Iteration: For each node, (i, j), $i \geq 1, j \geq 1$, of the grid, we compute the accumulated similarity cost:

$$H(i, j) = \max \begin{cases} H(i - 1, j - 1) + S(i, j), \\ H(i - k, j) - (1 + \frac{1}{3}k), & k \geq 1, \\ H(i, j - m) - (1 + \frac{1}{3}m), & m \geq 1. \\ 0 \end{cases} \qquad (7.5)$$

A closer inspection of Eq. (7.5) reveals that node (i, j) can be reached from the following predecessors:

- Node $(i - 1, j - 1)$, which refers to a diagonal transition. The local similarity between symbols $A(i)$ and $B(j)$ is added to the similarity that has been accumulated at node $(i - 1, j - 1)$.
- Nodes $(i - k, j)$, $k \geq 1$, which indicate vertical transitions. Such transitions imply a deletion (gap) of k symbols from sequence A (vertical axis) and introduce a gap penalty, $W_k = -(1 + \frac{1}{3}k)$, proportional to the number of symbols being deleted.

- Nodes $(i, j - m)$, $m \geq 1$ address horizontal transitions, m symbols long. The associated gap penalty is equal to $W_m = -(1 + \frac{1}{3}m)$ and suggests that m symbols need to be skipped from sequence B during the sequence alignment procedure.
- If all the above predecessors lead to a negative accumulated score, then $H(i, j) = 0$ and the predecessor of (i, j) is set equal to the fictitious node $(0, 0)$.

If, on the other hand, the accumulated score is positive, the best predecessor, $P_B(i, j)$, of (i, j) is the node that maximizes $H(i, j)$,

$$P_B(i, j) = \underset{(0,0),(i-1,j-1),(i-k,j):k\geq 1,(i,j-m):m\geq 1}{\arg\max} H(i, j) . \qquad (7.6)$$

Equations (7.5) and (7.6) are repeated for all nodes of the grid, starting from node $(1, 1)$ and moving from left to right (increasing index j) and from the bottom to the top of the grid (increasing index i). Concerning the first row and first column of the grid, the predecessor is always set to the fictitious node $(0, 0)$. All predecessors, $P_B(i, j)$, $i = 0, \ldots, M, j = 0, \ldots, N$, are stored in a separate $(M + 1) \times (N + 1)$ matrix to enable the final processing stage (the *backtracking* operation).

Backtracking: After $H(i, j)$ has been computed for all nodes, we select the node that corresponds to the maximum score and follow the chain of predecessors until we encounter a $(0, 0)$ node (the algorithm ensures that a $(0, 0)$ node will definitely be encountered). The resulting chain of nodes is the *best (optimal alignment) path*. The coordinates of the nodes of the best path reveal the subsequences of A and B, which have been optimally aligned, and indicate which symbol deletions have taken place. The backtracking procedure can be repeated to generate the second best alignment and so on. Care has to be taken so that each new backtracking operation ignores the nodes that have already participated in the preceding optimal paths.

We have implemented the standard S-W algorithm in the `smithWaterman()` function. To reproduce the example in the original paper ([99]), type:

```
[H,Bp_row,Bp_col,maxH,best_path]=smithWaterman('AAUGCCAUUGACGG', ...
                                   'CAGCCUCGCUUAG',1/3,1,0);
```

The first two arguments are molecular sequences and the third argument controls the gap penalty. The fourth argument is set to '1' because we want to have the output of the alignment procedure printed on the screen. If the last argument is set to '1', then a figure with the best path is also plotted. The printout of the function call is:

```
Similarity = 3.33

    G <-> G      (match)
    C <-> C      (match)
    C <-> C      (match)
    A <-          (deleted)
    U <-> U      (match)
    U <-> C      (replacement)
    G <-> G      (match)
```

It can be seen that $G - C - C - A - U - U - G$ has been aligned with $G - C - C - U - C - G$. The second subsequence is one symbol shorter, hence the deletion of symbol A. The deletion reduces the accumulated similarity score by $1 + \frac{1}{3}$. Furthermore, the mapping of U to C is equivalent to a symbol replacement that reduces the similarity score by $\frac{1}{3}$. This is why the alignment score is equal to $3.33 = 5 \times 1 - (1 + \frac{1}{3}) - \frac{1}{3}$. Instead of character strings, we can also use the smithWaterman() function to align any type of sequences consisting of discrete symbols. The next example uses integers. Type:

```
[H,Bp_row,Bp_col,maxH,best_path]=smithWaterman([1 −1 1 −1 1 1 2 2 1],...
                                [1 −1 1 −1 1 2 1 2],1/3,1,0);
```

In this case, the output on the screen is:

```
Similarity =   5.67

    1 <-> 1      (match)
   −1 <-> −1     (match)
    1 <-> 1      (match)
   −1 <-> −1     (match)
    1 <-> 1      (match)
      -> 2       (deleted)
    1 <-> 1      (match)
    2 <-> 2      (match)
```

7.2. HIDDEN MARKOV MODELING

A *Hidden Markov Model (HMM)* is a stochastic automaton that undergoes distinct transitions among states based on a set of probabilities. When the HMM arrives at a state, it emits an observation, which, in the general case, can be a multidimensional feature vector of discrete or continuous elements. The emission of observations is also governed by probabilistic rules.

During the design of an HMM it is of crucial importance to decide on the number of states and assign a meaningful stochastic interpretation to each state, so that the nature of the problem is reflected successfully on the structure of the HMM. These are not trivial issues in any sense and usually require a careful study of the stochastic nature of the phenomenon for which the HMM will be designed. For example, a phoneme in a speech recognition task can be modeled with a 3-state HMM, because, from a statistical perspective, the phoneme consists of three different parts (segments) and this is also reflected in the properties of most short-term feature sequences that can be extracted from the phoneme (e.g. the sequence of MFCCs). From a simplified point of view, the three parts refer to the beginning, steady state, and attenuation of the phoneme. As another example, in a speech/music/silence segmentation scenario, it can be useful to assign a class label to each state, leading to a 3-state HMM. As a general remark, the number of states is implicitly dictated by the interpretation of the most important stationarity changes in the signals of the problem at hand.

Another important issue has to do with the selection of the most appropriate features for the signals under study. For example, in speech recognition, it has been customary to use vectors of MFCCs as features and this has also turned out to be a more general trend in the field of audio processing. In the general case, the selected features should be able to reflect the temporal evolution of the signal on a short-term basis, without providing excessive detail. The designer of the HMM must be familiar with various audio features (see also Chapter 4) and needs to be able to experiment with their combinations. Usually, the resulting feature vectors are multidimensional and consist of continuous features. However, in order to provide a gentle introduction to the theory of HMMs, we focus on the case of discrete observations.

7.2.1. Definitions

Let $S_i, i = 1, \ldots, N$ be the set of HMM states and $b_j, j = 1, \ldots, M$ the discrete symbols of the alphabet of observations. The following three matrices represent the HMM, namely:

- The vector of *initial state probabilities*, $\underline{\pi}$. Element, $\underline{\pi}(i), 1, \ldots, N$ is the probability that the HMM starts its operation at the ith state.
- The *state transition matrix*, A. We define that $A(i, j) = p(j \mid i)$, i.e. $A(i, j)$ is the probability that the model jumps from state i to state j. Obviously, A is a $N \times N$ matrix and $\sum_{j=1}^{N} A(i, j) = 1, \forall i$.
- The matrix of *emission probabilities*, B. We define that $B(m, i) = p(b_m \mid i)$. This means that $B(m, i)$ is the probability that the mth symbol of the alphabet is emitted from the ith state. It follows that B is a $M \times N$ matrix and that $\sum_{m=1}^{M} B(m, i) = 1, \forall i$.

7.2.2. An Example

Consider a 3-state HMM that is capable of emitting two discrete symbols. The HMM is represented by the following matrices:

$$\underline{\pi} = [1\ 0\ 0]^T, A = \begin{bmatrix} 0.7 & 0.3 & 0 \\ 0 & 0.6 & 0.4 \\ 0 & 0 & 1 \end{bmatrix} \quad \text{and} \quad B = \begin{bmatrix} 0.8 & 0.4 & 0.1 \\ 0.2 & 0.6 & 0.9 \end{bmatrix}.$$

It can be observed that the HMM always starts at the first state and that matrix A is upper triangular. The latter implies that we are dealing with a *left-to-right* model because backward transitions do not exist, or, equivalently, a transition from a state always ends at a state with a higher index. This type of HMM has been very popular in the speech recognition field because it fits well with the temporal evolution of phonemes, the building blocks of speech utterances.

Now assume that we ask the following question: what is the probability, p_{sym}, that the sequence of states $\{S_1, S_2, S_2, S_3\}$ emits the sequence of symbols $\{x_1, x_1, x_2, x_2\}$? The answer is based on computing the joint probability of the following events:

$$p_{sym} = \underline{\pi}(1) \cdot p(x_1 \mid S_1) p(S_2 \mid S_1) \cdot p(x_1 \mid S_2) \cdot p(S_2 \mid S_2) \cdot$$
$$p(x_2 \mid S_2) \cdot p(S_3 \mid S_2) \cdot p(x_2 \mid S_3)$$
$$= 1 \cdot 0.8 \cdot 0.3 \cdot 0.4 \cdot 0.6 \cdot 0.6 \cdot 1 \cdot 0.9 = 0.0311.$$

7.2.3. Some Useful Questions

Given an HMM, $\mathcal{H}$, it is important to be able to answer the following questions, which are encountered in almost every application involving HMMs:

1. *Find the sequence of states that emits the sequence of observations* $X = \{x_1, x_2, \ldots, x_T\}$ *with the highest probability.* The answer to this question is given by the *Viterbi* algorithm, a dynamic programming algorithm, which is further described in Section 7.3. The respective probability is also known as the *Viterbi score*.
2. *Compute the sum of probabilities from all sequences that are capable of emitting* X. The result is known as the *recognition score* or *recognition probability* of the observation sequence given the HMM. To compute this score, we employ the *Baum-Welch* algorithm, which is presented in Section 7.4.
3. *Assign values to the parameters of the HMM.* So far, we assumed that the matrices (parameters) of the HMM are already known. In practice, they are the outcome of a *training stage*, which takes as input a set of observation sequences and determines the HMM parameters. The goal of the

training stage is to maximize a probabilistic criterion. If Viterbi training is adopted, the training algorithm attempts to maximize the sum of Viterbi scores of all observation sequences of the training set. If Baum-Welch training is used, then the goal is to maximize the respective sum of recognition probabilities. Section 7.5.1 presents an outline of the Viterbi training scheme. The reader is referred to [66] for a detailed treatment of the Baum-Welch training method.

After the HMM has been trained, it can be used both as a recognition machine and a generator of observation sequences, depending on the application context. For example, in a classification problem with four classes, we can train one HMM per class. When an unknown feature sequence is to be classified, it is assigned to the class which is represented by the HMM that has produced the highest recognition (or Viterbi) score. HMMs can be used for numerous other tasks, e.g. for audio tracking and segmentation purposes. It would not be an exaggeration to claim that HMM methodology is the driving technology in several machine learning fields, with speech recognition being an outstanding example.

7.3. THE VITERBI ALGORITHM

The goal of the Viterbi algorithm is to discover the optimal sequence of states given an HMM and a sequence of observations. Here, the term optimal is defined from a probabilistic perspective. In other words, the Viterbi algorithm seeks the sequence of HMM states that maximizes the joint probability of observations given the HMM.

The computational complexity of an exhaustive solution to the task is prohibitively high even for small-sized HMMs and short sequences of observations. For example, if a sequence is 30 observations long and an HMM consists of 5 states, the number of all possible state sequences is 5^{30}. It is, therefore, not a surprise that the algorithm resorts to dynamic programming principles to achieve its goal.

More specifically, it creates a cost grid (*trellis* diagram) by placing the states, $S(i)$, $i = 1, \ldots, N$ of the HMM on the vertical axis and the sequence of observations, $x(i)$, $i = 1, \ldots, T$, on the horizontal axis. Each node, (i, t), accumulates a probabilistic score, $\alpha_t(i)$, which is interpreted as the probability to reach state i, via an optimal sequence of states, after emitting the first t observations. The Viterbi algorithm computes the probability $\alpha_t(i)$ for every node of the trellis diagram, starting from the first column ($t = 1$). The maximum value at the last column is, therefore, the probability to emit the complete sequence of observations via an optimal sequence of states, which is subsequently extracted by means of a standard backtracking operation. We now provide a description of the steps of the Viterbi algorithm in the case of discrete observations. Remember that $\underline{\pi}$ is the vector of initial state probabilities, A is the state transition matrix, and B is the matrix of observation probabilities (Section 7.2).

Initialization: Assign values to the nodes of the first column of the grid:

$$\alpha_1(i) = \underline{\pi}(i)B(x(1), i), \, i = 1, \ldots, N.$$

It has been assumed that each symbol of the discrete alphabet has been given a positive identity. As a result, the feature sequence is actually a sequence of symbol identities and $B(x(1), i)$ stands for the probability that the ith state emits the symbol with identity $x(1)$.

Iteration: For the tth column, $t = 2, \ldots, T$, compute the probability $\alpha_i(t)$:

$$\alpha_i(t) = \max_{j=1,\ldots,N} \alpha_j(t - 1)A(j, i)B(x(t), i). \tag{7.7}$$

An interesting observation is that the computation of $\alpha_i(t)$ only takes into account the previous column of the grid $(t - 1)$. This is also known as the *first-order Markovian property*. The equation examines each node of the previous column as a potential predecessor of the node (i, t). To this end, $\alpha_j(t - 1), j = 1, \ldots, N$, is multiplied with the product $A(j, i)B(x(t), i)$, which is interpreted as the joint probability of the transition $j \to i$ and the emission of symbol $x(t)$ from state i. In the end, the winner is the transition that yields the maximum probabilistic score. Bellman's optimality principle [98] ensures that the resulting value, $\alpha_i(t)$, will represent the score of the optimal sequence of states that has ended with state i and has emitted the subsequence of observations until (including) the tth time instant. To enable backtracking, the coordinates of the best predecessor of node (i, t) are stored in a separate matrix, P_B, so that:

$$P_B(i, t) = \arg\max_{j=1,\ldots,N} \alpha_j(t - 1)A(j, i)B(x(t), i).$$

Termination: Detect the maximum value at the last column of the grid. This value is also known as the *Viterbi score* of the HMM for the given sequence of observations. The node that contains the Viterbi score serves to start a backtracking procedure that visits each column in reverse order until the first column is reached. The resulting chain of nodes reveals the best-state sequence (best-path).

We have implemented the Viterbi algorithm for discrete observations in the scaledViterbiDisObs() function, which is based on a *'scaling'* technique that deals with arithmetic stability issues due to the repeated multiplication of probabilities. The scaling method is based on the fact that if logarithms are used in Eq. (7.7), the products will be replaced by summations without affecting monotonicity. The scaled version of Eq. (7.7) is

$$\alpha_i(t) = \max_{j=1,\ldots,N} \log_{10}[\alpha_j(t-1)] + \log_{10}[A(j,i)] + \log_{10}[B(x(t),i)]. \quad (7.8)$$

To call the scaledViterbiDisObs() function, type:

```
[ViterbiScore,BestPath]=scaledViterbiDisObs(pinit,A,B,X);
```

where *pinit*, A, B, and X are the vector of initial probabilities, the transition matrix, the matrix of observation probabilities, and the sequence of observations (symbol identities), respectively. The output of the algorithm is the scaled version of the Viterbi score and the best-state sequence (vector of state identities).

7.4. THE BAUM-WELCH ALGORITHM

Instead of seeking the best-state sequence, it is often desirable to compute the sum of scores of all the sequences of states that are capable of generating a given sequence of observations. A small modification in Eq. (7.7) can provide the answer to this question. Specifically, if the maximization operator is replaced with a summation, Eq. (7.7) becomes:

$$\alpha_i(t) = \sum_{j=1}^{N} \alpha_j(t-1)A(j,i)B(x(t),i). \quad (7.9)$$

Equation (7.9) is computed for every node of the trellis diagram, following the node visiting scheme that was also adopted for the Viterbi algorithm. In the end, the values of the last column (Tth column) are summed to produce the *Baum-Welch score* (or *recognition probability*), r_{score}, of the HMM, given the sequence of observations:

$$r_{score} = \sum_{j=1}^{N} \alpha_j(T). \quad (7.10)$$

For this algorithm, it no longer makes sense to store the best predecessor of each node into a separate matrix and seek the best sequence of states, because every possible sequence of states contributes to the recognition score.

We have implemented a scaled version of the *Baum-Welch* algorithm in the `scaledBaumWelchDisObs()` function. The scaling technique is more complicated in this case, compared to the Viterbi method, and the reader is referred to [66] for a detailed description on the subject. To call the `scaledBaumWelchDisObs()` function, type:

```
[rScore]=scaledBaumWelchDisObs(pinit,A,B,X);
```

The function has preserved the input format of the `scaledViterbiDisObs()` function. Obviously, the recognition score is the only output argument in this case.

Further to the functions that operate on discrete observations, we also provide their counterparts for the case of continuous features. The respective functions have been implemented in the *scaledViterbiContObs()* and *scaledBaumWelchContObs()* m-files. The notable difference is the format of the argument that refers to emission of observations at each state. In the case of continuous features, it no longer makes sense to use a simple matrix format and we need to resort to a more complicated structure based on cell arrays, so that the pdf of each state is modeled successfully. A more detailed description is given in the next section in conjunction with the training procedure.

7.5. HMM TRAINING

The goal of the training stage is to 'learn' the parameters of the HMM. In the case of discrete observations, these are the vector of initial probabilities, π, the state transition matrix, A, and the matrix of emission probabilities, B. If continuous, one-dimensional observations are used, we will need to estimate a probability density function per state. For the more general case of continuous, multidimensional observations, we resort to multidimensional probability density functions to model the emission of observations at each state.

Irrespective of the nature of observations, the HMM training stage requires a *training dataset*, i.e. a set consisting of sequences of observations representative of the problem under study. The basic training steps are as follows:

1. Initialize the HMM parameters, either randomly or based on some prior knowledge of the problem at hand.
2. Compute the score (Baum-Welch or Viterbi), sc_i, for each sequence, $\mathbf{X}_i$, of the training set, where $i = 1, \ldots, L$ and L is the number of training sequences.

3. Compute the sum of scores, $sc_{total} = \sum_{i=1}^{L} sc_i$, that quantifies the extent to which the HMM has learned the problem. This is the quantity that the training stage seeks to maximize.
4. If sc_{total} remains practically unchanged between successive iterations, terminate the training stage, otherwise, proceed to the next step.
5. Update the HMM parameters. The procedure, which updates the weights, depends on the nature of the features and the type of score that has been adopted.
6. Repeat from Step 2.

HMM training is a machine learning procedure. As is often the case with learning algorithms, the initialization scheme plays an important role. For example, if a parameter is initialized to a zero, it will never change value. Another effect of poor initialization is that the training algorithm may get trapped at a local maximum. Furthermore, there are several issues related to the size and quality of the training set. For example, a common question is how many training sequences are needed, given the size (number of parameters) of an HMM. The tutorial in [66] provides pointers to several answers to these issues. In the next section, we present the basics of Viterbi training for sequences of discrete observations. The Viterbi training scheme was chosen because its simplicity allows for an easier understanding of the key concepts that underlie HMM training.

7.5.1. Viterbi Training

Assume that we have initialized an HMM by means of an appropriate initialization scheme. Viterbi training follows the aforementioned training outline and is based on the idea that if each training sequence is fed as input to the HMM, the Viterbi algorithm can be used to compute the respective score and best-state sequence.

7.5.1.1. Discrete Observations

The procedure, which updates the HMM parameters, is based on computing frequencies of occurrences of events following the inspection of the extracted best-state sequences. For example, consider that the following two sequences of observations and respective best-state sequences are available during the HMM training stage, where S_i, $i = 1, \ldots, 3$, are the HMM states and b_1, b_2 the two discrete symbols of the alphabet:

$$\{b_1 \; b_1 \; b_2 \; b_1 \; b_1 \; b_2 \; b_2\} \qquad \{b_2 \; b_2 \; b_1 \; b_1 \; b_2 \; b_2 \; b_2\},$$
$$\{S_1 \; S_1 \; S_2 \; S_2 \; S_3 \; S_3 \; S_3\} \qquad \{S_1 \; S_2 \; S_2 \; S_2 \; S_3 \; S_3 \; S_3\}.$$

To understand the update rule, let us focus on element $A(1, 1)$, which stands for the probability $p(S_1 \mid S_1)$, i.e. the probability that the model makes a self-transition to S_1. If we inspect the two best-state sequences, we understand that this particular self-transition has only occurred once. In addition, it can be observed that, in total, three transitions took place from S_1 to any state. The conditional probability, $p(S_1 \mid S_1)$, is therefore updated as follows:

$$\widehat{p}(S_1 \mid S_1) = \frac{num.\ of\ occurrences\ of\ S_1 \rightarrow S_1}{total\ num.\ of\ transitions\ from\ S_1} = \frac{1}{3}.$$

If this rationale is repeated for every element of matrix A, the updated matrix, $\widehat{A}$, is:

$$\widehat{A} = \begin{bmatrix} \frac{1}{3} & \frac{2}{3} & 0 \\ 0 & \frac{3}{5} & \frac{2}{5} \\ 0 & 0 & \frac{4}{4} \end{bmatrix}.$$

In the general case,

$$\widehat{A}(i, j) = \widehat{A}(S_j \mid S_i) = \frac{num.\ of\ occurrences\ of\ S_i \rightarrow S_j}{total\ num.\ of\ transitions\ from\ S_i},$$

We now turn our attention to matrix B. In this case, the update rule for element $B(i, j)$ is:

$$\widehat{B}(i, j) = \widehat{B}(b_i \mid S_j) = \frac{num.\ of\ times\ b_i\ was\ emitted\ from\ S_j}{total\ num.\ of\ symbols\ from\ S_j}.$$

In the above example, $\widehat{B}(2, 2) = \frac{2}{5}$.

Finally, the vector, $\underline{\pi}$, of initial probabilities is updated as:

$$\widehat{\pi}(i) = \frac{num.\ of\ times\ S_i\ was\ the\ first\ sequence}{total\ num.\ of\ sequences}.$$

In our example, $\widehat{\pi}(1) = \frac{2}{2} = 1$.

We provide an implementation of the Viterbi training scheme for the case of discrete observations in the `viterbiTrainingDo()` function. Before you call this function, you need to decide on the number of states, the length of the discrete alphabet, and wrap the training set in a cell array (*TrainingSet*, fifth input argument in the call below). Each cell contains a sequence of discrete observations (symbol identities). To call the function, type:

```
[pi_hat,A_hat,B_hat,probSum]=viterbiTrainingDo(pi_init,A_init, ...
                B_init, NumSymbols, TrainingSet, epochs, mindiff);
```

The first four input arguments serve to initialize the training procedure. The sixth input argument controls the maximum number of iterations (epochs) during which the training set will be processed until the algorithm terminates. The algorithm computes a sum of scores at the beginning of each iteration. If the absolute difference of this cumulative score is considered to be negligible between two successive iterations, i.e. the absolute difference is less than *mindiff* (last input argument), the algorithm terminates, even if the maximum number of epochs has not been reached. The first three output arguments refer to the trained HMM. The last argument (*probSum*) is a vector, the *i*th element of which is the sum of Viterbi scores in the beginning of the *i*th stage.

7.5.1.2. Continuous Observations

If we are dealing with continuous, one-dimensional (or multidimensional) feature sequences, the updating procedure for vector π and matrix A is unaltered. However, a different approach needs to be adopted for emission probabilities because a continuous feature space demands the use of probability density functions at each state. In this case, it is common to assume that each pdf is a multivariate Gaussian or a mixture of multivariate Gaussians. In practice, the latter is a very popular assumption, because a pdf can be well approximated by a mixture of a sufficient number of Gaussians to the expense of increasing the number of parameters to be estimated.

The m-file `viterbiTrainingMultiCo` provides an implementation of the Viterbi training scheme for the general case of continuous, multidimensional features, under the assumption that the density function at each state is Gaussian. The main idea is that we inspect the output of the Viterbi scoring mechanism and, for each state, we assemble the feature vectors that were emitted by the state. The resulting set of vectors is used as input to a Maximum Likelihood Estimator [5], which returns the mean vector and covariance matrix of the multivariate Gaussian. To call the `viterbiTrainingMultiCo` function, type:

```
[pi_hat,A_hat,B_hat_Gauss,probSum]=viterbiTrainingMultiCo(pi_init,A_init,...
        B_init_Gauss, TrainingSet, epochs, mindiff);
```

The *B_init_Gauss* input argument is a nested cell array, the length of which is equal to the number of states. The *i*th cell, *B_init_Gauss{i}*, is a vector of cells: *B_init_Gauss{i}{1}* and *B_init_Gauss{i}{2}* are the mean vector and covariance matrix of the pdf of the *i*th state. The *TrainingSet* input argument is also a cell array: each cell contains a multidimensional feature sequence. The remaining input arguments have the same functionality as in the `viterbiTrainingDo()` function. Concerning the output arguments, a notable difference is the *B_hat_Gauss* cell array, the format of which follows the *B_init_Gauss* input variable.

A version of the Viterbi scheme for the case of Gaussian mixtures (function `viterbiTrainingMultiCoMix()`) is also provided in the accompanying software library. The header of the `viterbiTrainingMultiCo Mix()` function has preserved the format of `viterbiTrainingMul tiCo()`, although in this case, the *B_init_Gauss* argument is more complicated to reflect the fact that we have adopted Gaussian mixtures. At the end of this chapter, we provide several exercises that will help you gain a better understanding of the Viterbi training schemes that have been discussed.

7.6. EXERCISES

1. (D5) This exercise combines techniques from several other chapters. Complete the following steps:
   - Record your voice 10 times creating 10 WAVE files. Name the files *a1.wav, a2.wav, . . . , a5.wav, b1.wav, b2.wav, . . . , b5.wav*. The *a{1-5}.wav* files contain your voice singing a short melody that you like. Files *b{1-5}.wav* contain a different melody. Place all files in a single folder.
   - Write a function, say *function1*, that loads a WAVE file and uses a silence detector to remove the silence/background noise from the beginning and end of the recording. Then, it uses a short-term feature extraction technique to generate a sequence of MFCCs from the file.
   - Write a second function, *function2*, that takes as input the path to a folder and calls *function1* for each WAVE file in the folder. The resulting feature sequences are grouped in a cell array, *F*.
   - Select in random a feature sequence (example) from the *F* array and match it against all the other sequences using the Itakura local path constraints. Mark the top three results. They correspond to the three lowest matching costs. How many of them come from the same melody with the random example? Does the lowest matching cost refer to the correct melody?
   - Repeat using the Sakoe-Chiba constraints.
2. (D4) In a way, the previous exercise designs a binary classifier of melodies. Repeat the experiment using the leave-one-out validation method and compute the classification accuracy. An error occurs whenever none of the two lowest matching scores corresponds to the correct melody.

3. (D4) This exercise is about the so-called endpoint constraints that can be used to enhance a Dynamic Time Warping scheme. Specifically, let T_x and T_r be the lengths of feature sequences **X** and **R**, respectively. Sequence **X** is placed on the vertical axis of the DTW cost grid. Modify the *dynamicTimeWarpingItakura()* function so that instead of demanding that the backtracking procedure starts from node (T_x, T_r), we examine the nodes $(T_x - K, T_r), \ldots, (T_x, T_r)$ and select the one that has accumulated the lowest matching cost. K is a user-defined parameter. Its value depends on how much silence/background noise we expect to encounter after the end of the useful signal. For example, if the short-term processing hop size is equal to 20 ms and we expect that, on average, due to segmentation errors, 200 ms of silence are expected to be encountered at the end of the useful part of the feature sequence, then, $K = \frac{200 \text{ ms}}{20 \text{ ms}} = 10$. This exercise assumes that the sequence that has been placed on the horizontal axis is 'clean.' If this is not the case, we may have to seek the lowest matching score in more than one column at the right end of the grid.

4. (D5) Further to the right endpoint constraints, modify the *dynamicTimeWarpingItakura()* function, so that it can also deal with silence/background noise in the beginning (left endpoint) of the signal that has been mapped on the vertical axis?

5. (D4) Load the `WindInstrumentPitch` mat-file from the accompanying library. It contains the cell array *Seqs*, with pitch-tracking sequences that have been extracted from a set of monophonic recordings of a wind instrument, the clarinet. Select at random a pitch sequence, say **X**, and concatenate, in random order, all the remaining sequences, to create one single, artificial sequence, **A**. Use the *smithWaterman()* function to detect the subsequence of **A** that is most similar to **X**. Before you feed the sequences to the *smithWaterman()* function, round each value to the closest integer. Use different colors to plot in the same figure sequence **X** and the pattern that has been returned by the *smithWaterman()* function. Repeat the experiment several times and comment on your results. *Hint: The best path reveals the endpoints of the subsequence that is most similar to the query (sequence **X**).*

6. (D4) Repeat the previous experimental setup. However, this time, quantize each pitch value to the nearest semitone frequency. For convenience, we assume that the frequencies, f_k, of the semitones of the chromatic scale span five octaves in total, starting from 55 Hz, i.e. they are given by the

equation:

$$f_k = f_0 \cdot 2^{\frac{k}{12}}, \quad f_0 = 55 \text{ Hz}, \quad k = 0, 1, \ldots, 59.$$

Have the results improved in comparison with the previous exercise, which was simply rounding each pitch value to the closest integer?

7. (D4) In this exercise, you are asked to implement a variant of the Smith-Waterman algorithm that can receive as input, continuous, multidimensional feature vectors. To this end, you will need to modify the similarity function, so that it can operate on multidimensional feature vectors. A similarity function, $S(i, j)$, that can be adopted is the cosine of the angle of two vectors, $\underline{x}$ and $\underline{y}$, which is defined as:

$$S(i, j) = \frac{\sum_{k=1}^{L} \underline{x}_k \underline{y}_k}{\sqrt{\sum_{k=1}^{L} \underline{x}_k^2} \sqrt{\sum_{k=1}^{L} \underline{y}_k^2}},$$

where L is the dimensionality of the feature space. By its definition, the proposed similarity measure is bounded in $[-1, 1]$. Negative values indicate dissimilarity. An interesting property of the cosine of the angle of two vectors is that it is invariant to scaling. A physical interpretation of scaling is that if the two vectors are multiplied by constants, the cosine of their angle is not affected.

Modify the *smithWaterman()* function, so that it can deal with continuous, multidimensional feature vectors, by adopting the aforementioned similarity function.

8. (D5) Create a MATLAB function that enhances the Viterbi algorithm. In its standard version, the algorithm examines the nodes $(i, t - 1)$, $i = 1, \ldots, N$, of the previous column of the grid to compute the score at node (i, t). Furthermore, the cost, $T[(i_1, t - 1) \rightarrow (i_2, t)]$, of the transition $(i_1, t - 1) \rightarrow (i_2, t)$ is equal to

$$T[(i_1, t - 1) \rightarrow (i_2, t)] = A(i_2 \mid i_1) B(b_t \mid i_2),$$

where b_t is the tth observation, assuming a discrete alphabet. In this enhanced version, K previous columns need to be searched. Furthermore, we depart from the world of discrete observations and assume that the feature sequence consists of continuous, multidimensional feature vectors. In addition, the cost of the transition, $(i_1, t - K) \rightarrow (i_2, t)$, is now defined as:

$$T[(i_1, t - K) \rightarrow (i_2, t)] = A(i_2 \mid i_1) f(\{b_{t-K+1}, b_{t-K+2}, \ldots, b_t\} \mid i_2).$$

We can see that the new transition cost is a function of the K feature vectors, $\{b_{t-K+1}, b_{t-K+2}, \ldots, b_t\}$. For example, if we work with a two-state HMM, where each state is interpreted as an audio class, (e.g. speech vs music), we can define that

$$f(\{b_{t-K+1}, b_{t-K+2}, \ldots, b_t\} \mid i_2) \equiv P(\{b_{t-K+1}, b_{t-K+2}, \ldots, b_t\} \mid i_2),$$

which is the posterior probability that the K feature vectors belong to the i_2 class (speech or music). If, in addition, the transitions among states are equiprobable, the proposed enhanced Viterbi scheme solves the segmentation problem that was also examined in Section 6.1.2 from a cost-function optimization perspective.

In this exercise, your implementation, say *EnhancedViterbi()*, will receive as input a vector of initial probabilities, a transition matrix, the K parameter, a feature sequence, and the name of the m-file in which function f is implemented. The output is the enhancedViterbi score and the best-state sequence.

9. (D5) Develop a very simple, HMM-based classifier of melodies. Load the `WindInstrumentPitch` mat-file. Further to the *Seqs* array, a second cell array, *Tags* is also loaded into memory. It contains the names of the clarinet files from which the pitch sequences were extracted. Naming convention requires that all filenames that start with the same digit refer to variations of the same melody. For example, *clarinet11e.wav* and *clarinet11i.wav* are variations of the melody in the *clarinetModel1.wav* file. You will now need to go through the following steps:

   (a) Plot the sequence that refers to the *clarinetmodel2.wav* file. The respective tag in the *Tags* cell array is 'clarinetmodel2.wav.' It can be observed that it is a sequence of five notes.

   (b) Round each value to the closest integer.

   (c) Create a left-to-right, 5-state HMM, where each state corresponds to a note. Each state is only linked with itself and the next state in the chain. The initial probability of the first state is equal to 1.

   (d) Each state emits discrete observations. The highest emission probabilities stem from a narrow range of integers around the frequency of the note. The range of integers is a user-defined parameter and its physical meaning is our tolerance with respect to the pitch frequency of the note. All frequencies outside the narrow interval of tolerance are assigned very low probabilities, in the order of 0.001.

   (e) Create a second HMM for the melody in the *clarinetmodel1.wav* file. The respective tag in the *Tags* cell array is 'clarinetmodel1.wav.'

(f) After you construct the two HMMs and set their parameters manually (no training is involved), select a variation of one of the two melodies, e.g. the *clarinet11e.wav* sequence. Feed the sequence to both HMMs and compare the Viterbi scores.

(g) Repeat for all variations and compute the classification accuracy.

10. (D5) This exercise simulates an audio tracking scenario:

   (a) Load the *musicSmallData* mat-file. It contains feature sequences from 15 audio recordings.

   (b) Use only the MFCCs and discard the rest of the features.

   (c) Select the sequence of MFCCs from one recording and use it as the example to be searched.

   (d) Concatenate all sequences of MFCCs (including the example) to form a single sequence, Y.

   (e) Distort sequence Y as follows: every M feature vectors (M is a user parameter), select one vector in random. Generate a random number, r, in the interval $[0, 1]$. If $r > 0.65$, repeat the selected feature vector once, i.e. insert a copy of the vector next to the existing one. If $0.3 \leq r \leq 0.65$, add random noise to each MFCC coefficient of the vector. The intensity of the noise should not exceed a threshold that is equal to ($0.1 \times$ the magnitude of the coefficient). Finally, if $r < 0.3$, perform a random permutation of the M feature vectors.

   (f) Use the Itakura and Sakoe-Chiba local path constraints to detect the example in the distorted version of sequence Y. It is important to remember that, for this task, backtracking can start from any node of the last G columns (G is a user-defined parameter) and Y is placed on the vertical axis.

   (g) Repeat the detection operation using the Smith-Waterman algorithm. Use the cosine of the angle of two feature vectors as a local similarity function.

   (h) Experiment with different examples and comment on the results.

PART Three

Other Issues

Other Issues

Music Information Retrieval

Contents

During recent years, the wide distribution of music content via numerous channels has raised the need for the development of computational tools for the analysis, summarization, classification, indexing, and recommendation of music data. The interdisciplinary field of Music Information Retrieval (MIR) has been making efforts for over a decade to provide reliable solutions for a variety or music related tasks [105,106], bringing together professionals from the fields of signal processing, musicology, pattern recognition and psychology, to name but a few. The following is a brief outline of some popular MIR application fields:

- *Musical genre classification.* As already discussed in Chapter 5, musical genre classification refers to the task of classifying an unknown music track to a set of pre-defined musical genres (classes), based on the analysis of the audio content [14]. Despite the inherent difficulties in providing a rigorous definition of the term 'music genre', several studies have shown that it is possible to derive solutions that exhibit acceptable performance in the context of genres of popular music [14,107–109].

- *Music identification.* This involves identifying a musical excerpt, using a system that is capable of matching the excerpt against a database of music recordings. A popular mobile application that falls in to this category is the Shazam mobile app [110]. A key challenge that music identification systems have to deal with is that the example can be a seriously distorted signal due to recording conditions (noisy environment, poor quality recorder). Furthermore, they need to exhibit low response times (of the order of a few seconds) and the database of reference music tracks has to be as extensive as possible.

- *Query-by-humming (QBH).* This is the process of [111,112] identifying a sung or whistled excerpt from a melody or tune. Initially, the excerpt is analyzed to extract a pitch-tracking sequence, which is then matched against a database of melodies that have been stored in symbolic format (e.g. MIDI). Query-by-humming systems rely heavily on DTW (dynamic time warping) techniques to achieve their goal and fall in to the broader category of audio alignment techniques. QBH systems must be robust with respect to recording conditions and the difficulties that most users face when producing a melody correctly (accuracy of notes and tempo). Recently, several mobile applications have been made available that provide solutions to variants of the basic QBH task.

- *Visualization.* Large music archives usually provide text-based indexing and retrieval mechanisms that rely on manually entered text metadata about the artist, the musical genre, and so on. A useful complementary approach is to assist the browsing experience of the user by means of visualizing the music content in a 2D or 3D space of computer graphics, so that similar pieces of music are closer to each other in the resulting visual representation [113]. To this end, several methods have been adopted to reduce the dimensionality of the feature space, including Self-Organizing Maps (SOMs) [114–118], regression techniques [119], and the Isomap algorithm [120], to name but a few. The visualization systems must also be able to account for the preferences of the users, i.e. incorporate personalization mechanisms [115,119]. A recent trend is to use emotional representations of music to complement the browsing experience of users [121–123].

- *Recommender systems.* In general, the goal of recommender systems is to predict human preferences regarding items of interest, such as books, music, and movies. Popular music recommendation systems are Pandora, LastFm, and iTunes Genius. In general, we can distinguish two types of recommender systems: (a) collaborative filtering recommenders, which predict the behavior of users based on decisions taken by other users who are active in the same context, and (b) content-based systems, which compute similarities based on a blend of features, including features that are automatically extracted from the music content [124–126] and features that refer to manual annotations.

- *Automatic music transcription.* In the general case, it is the task of transforming a polyphonic music signal into a symbolic representation (e.g. MIDI) [127,128]. The simplest case of automatic music transcription refers to monophonic signals, i.e. music signals produced by one instrument playing one note at a time. A harder task is the transcription of the predominant melody of a music recording, i.e. the melody that has a leading role in the orchestration of the music track (usually, the singing voice). Music transcription relies heavily on pitch tracking and multipitch analysis models. Lately, source separation techniques have been employed as a preprocessing stage in music transcription systems.

- *Music thumbnailing.* This is the procedure of extracting the most representative part of a music recording [31]. In popular music, this is usually the chorus of the music track. The related techniques usually exploit the properties of the self-similarity matrix of the recording. A harder task is that of *music summarization*, where the goal is to perform the structural analysis of the music recording [129,130] and present the results in an appropriately encoded (possibly graphical) representation.

- *Cover song identification.* The term 'cover' refers to versions of a music recording, which may differ from an original track with respect to instrumentation, harmony, rhythm, performer, or even the basic structure [100]. The related usually rely on sequence alignment algorithms and can be very useful in the context of monitoring legal rights and detecting music plagiarism.

- *Music Meter/Tempo Induction, and Beat Tracking.* These are three related MIR tasks that have traditionally attracted significant research attention [131–133]. The computational study of rhythm and the extraction of related features can provide content-based metadata that can enhance the performance of any system in the aforementioned categories (e.g. [134]).

Throughout the rest of this chapter we provide implementations of selected MIR tasks, so that the reader can gain a better understanding of the field, while also making use of the techniques and respective functions that were presented in previous chapters of the book. The selected tasks and respective implementations aim at providing the beginner in the field of MIR with the material that helps them take their first steps in the field but by no means constitutes an extensive coverage of the area.

8.1. MUSIC THUMBNAILING

Automated structural analysis of music recordings has attracted significant research interest over the last few years due to the need for fast browsing techniques and efficient content-based music retrieval and indexing mechanisms in large music collections. In the literature, variations of the problem of structural analysis [129,130,135,136] are encountered under different names, including music summarization, repeated (or repeating) pattern finding, and music (audio) thumbnailing. In this section, we use the term *'music thumbnailing'* for the task of detecting instances of the most representative part of a music recording. We define an excerpt of music as representative if it is repeated at least once during the music recording and if it is of sufficient length (e.g. 30 s or longer). In the case of popular music, thumbnails are expected to coincide with the chorus of the music track, which is usually the easiest part to memorize. Music thumbnailing has a direct commercial impact because it is very common among music vendors on the Internet to allow visitors to listen to the most representative part of a music recording before the actual purchase takes place. Furthermore, the increasing distribution of music content via rapidly growing video sharing sites has also highlighted the need for efficient means to organize the available music content, so that the browsing experience becomes less time-consuming and more effective.

A large body of music thumbnailing techniques has adopted the *Self-Similarity Matrix (SSM)* of the recording as an intermediate signal representation [137]. This is also the case with the thumbnailing method that we are presenting here as a variant of the work in [31]. To compute the SSM of a music track, we first extract a sequence of short-term chroma vectors from the music recording, using the `stFeatureExtraction()` function that was introduced in Chapter 4.

```
% Short-term Feature Extraction for all features
Features = stFeatureExtraction(x, fs, winSize, winStep);
% Select the chroma vector from each frame
Chromagram = Features(23+1:23+12, :); % feature starts at row 23+1
```

To reduce the computational burden, we recommend that $winSize = winStep = 0.25$ s.

We then compute the pairwise similarity for every pair of vectors, $(\underline{x}_i, \underline{x}_j)$, $i, j = 1, 2, \ldots, M$, where M is the length of the feature sequence. The adopted similarity function is the correlation of the two vectors, defined as:

$$S(i, j) = \sum_{r=1}^{12} \underline{x}_i(r) \cdot \underline{x}_j(r), \quad i, j = 1, 2, \ldots, M \tag{8.1}$$

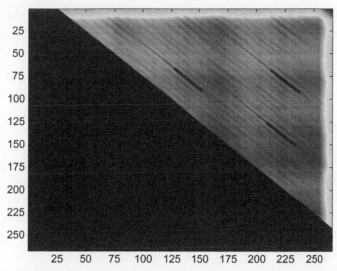

Figure 8.1 Self-similarity matrix for the track 'True Faith' by the band New Order. Detected thumbnails are shown with superimposed lines. For example, a pair of thumbnails occurs at segments with endpoints (in seconds) (66, 91) and (127, 152). This corresponds to two repetitions of the the song's chorus.

The resulting similarity values form the $(M \times M)$ self-similarity matrix S. Due to the fact that $S(i, j) = S(j, i)$, the SSM is symmetric with respect to its main diagonal, so we only need to focus on the upper triangle (where $i < j$) during the remaining steps of the thumbnailing scheme. Figure 8.1 presents the self-similarity matrix of a popular music track.

The next processing stage bears its origins in the field of image analysis. Specifically, a diagonal mask is defined and applied as a *moving average filter* on the values of the SSM. The adopted mask is the $(K \times K)$ identity matrix, $I_{K \times K}$, where K is a user parameter that defines the desirable thumbnail length. By its definition, the mask has 1s on the main diagonal and 0s in all other elements. As an example, if the desirable thumbnail length is 30 s and the short-term processing step was previously set equal to 0.25 s, then $K = \frac{30}{0.25} = 120$. Due to the fact that the mask is centered at each element of matrix S, we prefer odd values for K, so we would actually use $K = 121$ in this case.

The result of the moving average operation is a new matrix, $\widehat{S}$, defined as

$$\widehat{S}(i, j) = \sum_{k=-\frac{K-1}{2}}^{+\frac{K-1}{2}} \sum_{l=-\frac{K-1}{2}}^{+\frac{K-1}{2}} S(i-k, j-l) \cdot I\left(k + \frac{K-1}{2}, l + \frac{K-1}{2}\right). \quad (8.2)$$

After the masking operation has been completed, we find the highest value of matrix $\widehat{S}$. If this value occurs at element $\widehat{S}(a, b)$, the respective pair of thumbnails consists of the feature sequences with indices

$$\left\{ a - \frac{K-1}{2}, \ldots, a + \frac{K-1}{2} \right\}, \left\{ b - \frac{K-1}{2}, \ldots, b + \frac{K-1}{2} \right\}.$$

In our implementation, the selection of highest value in matrix $\widehat{S}$ has been extended to cover (at most) the top three similarities. As a result, at most, three pairs of thumbnails will be extracted. In Figure 8.1, the detected pairs have been superimposed with lines on the SSM of the figure.

The `musicThumbnailing()` function implements our version of the music thumbnailing scheme. It returns the endpoints (in seconds) of one up to three pairs. The following MATLAB code demonstrates how to use the function for a music track that is stored in a WAVE file:

```
songPath = 'song.wav';
% read the WAV data:
[x,fs] = wavread(songPath);
% get the thumbnail limits:
sWin = 0.25; sStep = 0.25; thumbnailSize = 30.0;
t = musicThumbnailing(x, fs, sWin, sStep, thumbnailSize, 3, 0);
% read the 1st couple of thumbnails:
[x1, fs] = wavread(songPath, round([t(1,1) t(1,2)] * fs));
[x2, fs] = wavread(songPath, round([t(1,3) t(1,4)] * fs));
% play the two thumbnails:
fprintf('Playing the 1st segment\n');
sound(x1, fs);
fprintf('Press any key to play the 2nd segment...\n');pause
fprintf('Playing the 2nd segment\n');
sound(x2, fs);
```

8.2. MUSIC METER AND TEMPO INDUCTION

In this section, we present one more application of self-similarity analysis: a method that is capable of extracting the music meter and tempo of a music recording. The proposed scheme is a variant of the work in [138], which is based on the observation that the rhythmic characteristics of a music recording manifest themselves as inherent signal periodicities that can be extracted by processing the diagonals of the Self-Similarity Matrix (SSM).

The stages of the method are as follows:
1. The music signal is long-term segmented and the SSM of each long-term segment is generated. The short-term feature that we use is the set of standard MFCCs, although this is not a restrictive choice and the reader is encouraged to experiment with different sets of features. The Euclidean

distance function is used to compute pairwise dissimilarities, so, strictly speaking, we are generating a dissimilarity matrix per long-term segment.

2. The mean value of each SSM diagonal is computed. A low mean value indicates a strong periodicity, with a period, measured in frames, equal to the index of the diagonal.

3. The sequence of mean values is treated as a signal, B. An approximation, D_2, of the second derivative of this signal is computed and its maxima are detected.

4. The detected local maxima provide a measure of sharpness to the corresponding local minima of B (signal periodicities).

5. The local maxima are examined in pairs to determine if they fulfill certain criteria. The two main criteria are that the round-off error of the ratio of the indices of the respective diagonals should be as close to zero as possible (an appropriate threshold is used) and that the sum of heights of the peaks should be as high as possible. The best pair of local maxima, with respect to the selected criteria, is finally selected as the winner. The periodicities of the pair determine the tempo and music meter of the recording. If none of the pairs meets the criteria, the long-term segment returns a null value.

For the short-term feature extraction stage, recommended values for the length and step of the moving window are 0.1 and 0.005, respectively. If each long-term segment is approximately 10 s long, 2000 vectors of MFCCs are generated from each long-term window and the dimensions of the resulting SSM are (2000 × 2000). This high computational load is due to the small hop size of the short-term window. However, this is the price we pay if we want to be able to measure the signal periodicities with sufficient accuracy.

To compute an approximation, $\mathbf{D}$, of the second-order derivative of $\mathbf{B}$, we first approximate the first-order derivative, $\mathbf{D}_1$, and use the result to estimate the second-order derivative:

$$D_1(k) = \sum_{l=-3}^{3} B(k+l), \quad k = 1, \ldots, L \ \text{and}$$

$$D_2(k) = \sum_{l=-3}^{3} D_1(k+l), \quad k = 1, \ldots, L, \tag{8.3}$$

where L is the maximum lag (diagonal index) for which $D_2(k)$ is computed. Assuming that, in popular music slow music meters can be up to 4 s long, the maximum lag, L, for computing $\mathbf{D}_2$ is $L = \frac{4}{0.005} = 800$. Figure 8.2 presents the first 500 elements of sequence $\mathbf{D}_2$ for a 10 s excerpt from a relatively fast Rumba dance taken from the Ballroom Dataset [131]. The three highest

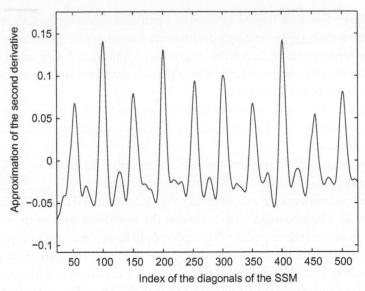

Figure 8.2 Approximation of the second derivative, D_2 of sequence **B**.

peaks occur at lags 100, 200, and 400. The respective signal periodicities are $100 \cdot 0.005 = 0.5$ s, $200 \cdot 0.005 = 1$ s, and $400 \cdot 0.005 = 2$ s, which correspond to the durations of the beat $\left(\frac{1}{4}\right)$, two beats $\left(\frac{2}{4}\right)$, and the music meter $\left(\frac{4}{4}\right)$.

After examining in pairs all peaks that exceed zero, the algorithm will eventually determine that the best pair of lags is (100, 400). The equivalent of the first lag in beats per minute (bpm) is $\frac{60}{100 \cdot 0.005} = \simeq 120$ (bpm). Due to the fact that $\frac{400}{100} = 4$ and 120 bpm is a typical quarter–note duration in popular music, for this particular segment, the method will return that the beat is 120 bpm and that the music meter is $\frac{4}{4}$.

We provide an implementation of this algorithm in the `musicMeter TempoInduction()` function. To call this function, you first need to load a signal (e.g. from a WAVE file) and set values for the parameters of the method. Type:

```
% Load the wav file. We use a file form the Ballroom Dataset.
% The file is assumed to reside in the working directory.
[x,fs,bits]=wavread('Albums-GloriaEstefan_MiTierra-04.wav'); %
lwLength=10; lwStep=10;
stLength=0.1; stStep=0.005;
roundoffError=0.15;
[Tempo, meterNum, meterDenom] = musicMeterTempoInduction(...
            x,fs,lwLength, lwStep, stLength , stStep, roundoffError, 1);
```

For a detailed explanation of the header of the function, type *help music MeterTempoInduction*. If the last input argument is set to 1, the function prints its output on the screen. If you are interested in a very detailed scanning of the signal, set the long-term step, lwStep, equal to 1. Note that the implementation has not been optimized with respect to response times or memory consumption, so you are advised to experiment with relatively short music recordings, e.g. 30 s long.

8.3. MUSIC CONTENT VISUALIZATION

Automatic, content-based visualization of large music collections can play an important, assisting role during the analysis of music content, because it can provide complementary representations of the music data. The goal of visualization systems is to analyze the audio content and represent a collection of music tracks in a 2D or 3D space, so that similar tracks are located closer to each other. In general, this can be achieved with dimensionality reduction methods which project the initial (high-dimensional) feature representation of the music data to low-dimensional spaces which can be easily visualized, while preserving the structure of the high-dimensional space. Dimensionality reduction approaches can be:

- *Unsupervised*: The dimensionality reduction of the feature space does not rely on prior knowledge, e.g. labeled data, because such information is unavailable.
- *Supervised and semi-supervised*: The techniques of this category exploit user-driven, prior knowledge of the problem at hand. For example, we may know track X is similar to track Y, so they have to be neighbors in the feature space, or that two feature vectors from the same music track have to be connected with a must-link constraint during a graph construction process.

In this section we provide brief descriptions of some widely used dimensionality reduction techniques that can be adopted in a content visualization context. In particular, we will present the methods of (a) *random projection*, (b) principal component analysis (PCA), (c) linear discriminant analysis, and (d) self-organizing maps. Other techniques that have been used for music content visualization are presented in [139], where multidimensional scaling (MDS) is adopted, and in [120], where the Isomap approach is used. Some of the methods have been left as exercises for the reader interested in delving deeper into the subject.

8.3.1. Generation of the Initial Feature Space

Before proceeding, let us first describe how to extract the initial feature space from a collection of music data based on the software library of this book.

This can be achieved with function featureExtractionDir(), which extracts mid–term feature statistics for a list of WAVE files stored in a given folder. For example, type:

```
[FeaturesDir, FileNames] = featureExtractionDir('musicFolder', ...
    0.020, 0.020, 10.0, 5.0, {'mean','std'});;
```

The function returns: (a) a cell array, where each cell contains a feature matrix, and (b) a cell array that contains the full path of each WAVE file of the input folder. Each element of the FeaturesDir cell array is a matrix of mid–term statistics that represents the respective music track. In order to represent the whole track using a single feature vector, this matrix is averaged over the mid–term vectors. This long–term averaging process is executed in functions musicVisualizationDemo() and musicVisualizationDemoSOM(), which are described later in this section. The reason that feature ExtractionDir() does not return long–term averages (which are, after all, the final descriptors of each music track) but returns mid–term feature vectors instead, is that some of the visualization methods that will be described later use mid–term statistics to estimate the transformation that projects the long–term averages to the 2D subspace.

The featureExtractionDir() function is executed for two directories: (a) the first one is a collection of MP3 files (almost 300 in total) that cover a wide area of the pop–rock genre, and (b) the second one is a smaller collection of the same genre styles. The feature matrices, along with the song tiles, are stored in the musicLargeData.mat and musicSmallData.mat files. These are then used by the proposed visualization techniques.

Note: The featureExtractionDir() function performs a mid–term feature extraction operation on a set of WAVE files, which are stored in the given folder. As it is more likely that music is stored in MP3 format, one may first need to use function mp3toWavDIR(), which transcodes all MP3 files in a folder in to their WAVE counterparts (see Section 2.4.2).

8.3.2. Linear Dimensionality Reduction

We start with three basic linear dimensionality reduction approaches, one of which is supervised. These methods extract a lower dimensionality space as a continuous linear combination of the initial feature space, as opposed to the

SOM approach (described in the next section), which operates in a discrete mode.

8.3.2.1. Random Projection

A straightforward solution to projecting a high-dimensional feature space to one of lower dimension, is by using a random matrix. Despite its simplicity, random projection has proved to be a useful technique, especially in the field of text retrieval, where the need to avoid excessive computational burden is of paramount importance [140]. Let M data samples be arranged columnwise in matrix $X_{D \times M}$, where D is the dimensionality of the feature space. The projection to the space of lower dimension is defined as:

$$Y_{d \times M} = R_{d \times D} X_{D \times M} \qquad (8.4)$$

where d is the new dimension and $R_{d \times D}$ is the random projection matrix. Equation (8.4) is also encountered in other linear projection methods; in the case of random projection, R is randomly generated. Note, that before applying random projection, a feature normalization step (to zero mean and unity standard deviation) is required, in order to remove the bias due to features with higher values.

8.3.2.2. Principal Component Analysis

Principal component analysis (PCA) is a widely adopted dimensionality reduction method aimed at reducing the dimensionality of the feature space while preserving as much 'data variance' (of the initial space) as possible [141,142]. In general, PCA computes the eigenvalue decomposition of an estimate of the covariance matrix of the data and uses the most important eigenvectors to project the feature space to a space of lower dimension. In other words, the physical meaning of the functionality of the PCA method is that it projects to a subspace that maintains most of the variability of the data.

PCA is implemented in the `princomp()` function of MATLAB's Statistics Toolbox. `princomp()`, accepts one single input argument, the feature matrix, and returns a matrix with the principal components. The first d columns of that matrix, which stand for the most dominant components, are used to project the initial feature space to the reduced d-dimensional space. From an implementation perspective, if `FeaturesDir()` is the cell array of mid-term feature statistics (output of `featureExtractionDir()`), then the PCA projection matrix is generated as follows:

```
X  = [];
for i=1:length(FeaturesDir)
```

```
    X = [X mean(FeaturesDir{i}, 2)]; % long-term averaging
end
X = X';
[COEFF, SCORE] = princomp(X);
% keep the coefficients of the
% two most dominant components eigenvectors:
V = COEFF(:, 1:2);
```

8.3.2.3. Fisher Linear Discriminant Analysis

PCA is an unsupervised method; it seeks a linear projection that preserves the variance of the samples, without using class information. However, we are often interested in maintaining the association among features and class labels, when such information is available. The Fisher linear discriminant analysis method (FLD, also known as LDA) is another linear dimensionality reduction technique that exploits the idea of embedding supervised knowledge in the procedure. This means that, in order to compute the projection matrix, we need to know the mappings of feature vectors in the initial feature space to predefined class labels. The basic idea behind LDA is that, in the lower-dimensional space, the centroids of the classes must be far from each other, while the variance within each class must be small [143]. Therefore, *LDA seeks the subspace that maximizes discrimination among classes, whereas PCA seeks the subspace that preserves most the variance among the feature vectors.*

The FLD method is not included in MATLAB. We have selected the fld() function, which is available on the Mathworks File Exchange website.[1] This function accepts the following arguments: a $X_{D \times M}$ feature matrix, a $L_{M \times 1}$ label matrix and the desired number of dimensions, d. It returns the optimal Fisher orthonormal matrix ($V_{d \times D}$), which is used to project the initial feature set to the new dimensionality.

The obvious problem with the FLD approach, in the context of a data visualization application, is that it *needs supervised knowledge* to function properly. So, the question is how we can obtain mappings of feature vectors to class labels. We adopt the following rationale to solve this problem:

- For each music track, extract mid-term feature statistics.
- Map the feature vectors from each music track to *the same class label*, i.e. one separate class label for each music track is created.
- Feed the FLD algorithm with the mid-term statistics and respective class labels (song IDs).

---

[1] Mathworks File Exchange, Fischer Linear Dicriminant Analysis, by Sergios Petridis: http://www.mathworks.com/matlabcentral/fileexchange/38950-fischer-linear-dicriminant-analysis.

In this way, the FLD method will try to generate a linear transform of the original data that minimizes the variance between the feature vectors (mid-term statistics) of the same class (music track), while maximizing the distances between the means of the classes. This trick enables us to feed the FLD algorithm with a type of supervised knowledge that stems from the fact that feature vectors from the same musical track must share the same class label. The extracted orthonormal matrix is used to project the long-term averaged feature statistics (one long-term feature vector for each music track). The following code demonstrates how to generate the FLD orthonormal matrix using the steps described above (again, FeaturesDir is a cell array of mid-term feature statistics):

```
X   = []; L = [];
for i=1:length(FeaturesDir)
  X = [X FeaturesDir{i}];
  L = [L; i*ones(size(FeaturesDir{i}, 2), 1)];
end
X = X';
[V, eigvalueSum] = fld(X, L, 2);
```

8.3.2.4. Visualization Examples

The three linear dimensionality reduction approaches that were described above (random projection, PCA, and LDA) have been implemented in the musicVisualizationDemo() function of the library. The function takes as input (a) the cell array of feature matrices, (b) the cell array of WAVE file paths, and (c) which dimensionality reduction method to use (0 for random projection, 1 for PCA, and 2 for LDA). The first two arguments are generated by featureExtractionDir(), as described in the beginning of Section 8.3.

We have demonstrated the results of the linear dimensionality reduction methods using the dataset stored in the musicSmallData.mat file, because the content of the large dataset would be a lot harder to illustrate. The musicSmallData.mat dataset is a subset of the larger one and consists of 40 audio tracks from 4 different artists. Two of the artists can be labeled as punk rock Bad Religion and NOFX and the other two as Synthpop (New Order and Depeche Mode). The code that performs the visualization of this dataset using all three methods (random projection, PCA, and LDA) is as follows:

```
load(['..' filesep 'data' filesep 'musicSmallData'])
musicVisualizationDemo(FeaturesDir, FileNames, 0); % Random projection
musicVisualizationDemo(FeaturesDir, FileNames, 1); % PCA
musicVisualizationDemo(FeaturesDir, FileNames, 2); % LDA
```

> *Note:* `musicVisualizationDemo()` also gives the user the ability to listen to the music tracks using the mouse on the 2D plane. However, this functionality requires the audio data (given by the paths stored in the `FileNames` variable). Therefore, the playback functionality can only be available for your own datasets and not for the examples of the `musicLargeData.mat` and `musicSmallData.mat` files (the software library does not provide the raw audio data).

Figure 8.3 shows the visualization results for all three linear projection methods. It can be seen that the LDA-generated visualization achieves higher discrimination between artists and genres (the songs of New Order are on the same plane as the Depeche Mode tracks). The random projection method confuses dissimilar tracks in many cases.

8.3.3. Self-Organizing Maps

Self-Organizing Maps (SOMs) provide a useful approach to data visualization. They are capable of generating 2D quantized (discretized) representations of an initial feature space [114–116]. SOMs are neural networks trained in an unsupervised mode in order to produce the desired data organizing map [144]. An important difference among SOMs and typical neural networks is that each node (neuron) of a SOM has a specific position in a defined topology of nodes. In other words, SOMs implement a projection of the initial high-dimensional feature space to a lower-dimensional *grid* of nodes (neurons).

In MATLAB, SOMs have been implemented in the Neural Network Toolbox. To map a feature space to a 2D SOM, we need to use the `self orgmap()`, `train()`, and `net()` functions. The following code demonstrates how this is achieved, given a matrix, X, that contains the vectors of the initial feature space:

```
nGrid = 5;
net = selforgmap([nGrid nGrid], 500, 3, 'gridtop');
net = train(net, X);
y = net(X);
classes = vec2ind(y);
```

Parameter `nGrid` stands for the dimensions of the grid of neurons (5×5). Variable `classes` contain the class label for each input sample. Note that the example adopted the `gridtop` topology, so class label 1 corresponds to the neuron at position (0,0), class label 2 to position (1,0), and so on. The `gridtop`

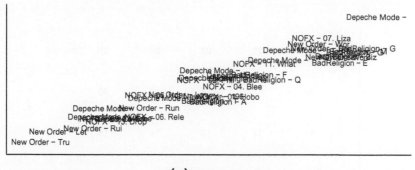

(a) Random projection

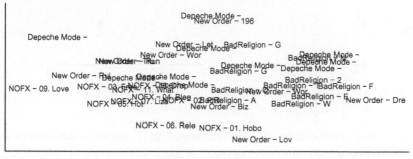

(b) PCA

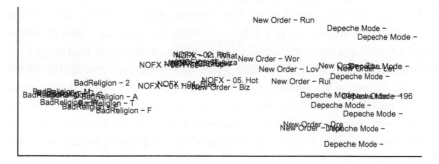

(c) LDA

Figure 8.3 Visualization results for the three linear dimensionality reduction approaches, applied on the `musicSmallData.mat` dataset. Note that the length of the track titles has been truncated for the sake of clarity.

topology of the SOM is shown in Figure 8.4, which is for a 5 × 5 grid. Other MATLAB-supported grids are the `hextop` and `randtop` topologies.

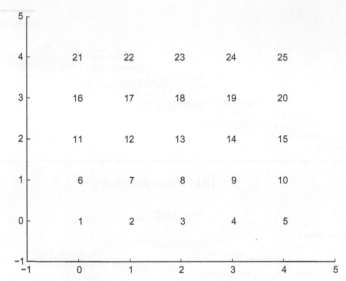

Figure 8.4 `gridtop` topology (5 × 5). Numbers indicate neuron IDs.

Function musicVisualizationDemoSOM() demonstrates how to use SOMs to map a high-dimensional space of audio features to a 2D plane. Note that the first two input arguments of this function are similar to the arguments of the musicVisualizationDemo() function. The third input argument is the size of the grid and the fourth is the type of method: 0 for direct SOM dimensionality reduction and 1 for executing the LDA as a preprocessing step.[2] Overall, the musicVisualizationDemoSOM() function maps each feature vector of the initial feature space to a node (neuron) of the SOM grid. *In the end of the training phase, each node of the grid contains a set of music tracks.* The number of tracks that have been assigned to each node is plotted in the respective area of the grid and when the user clicks on a node, the respective music track titles are visualized in a separate plotting area (see Figure 8.5).

We have used the audio features in the musicLargeData.mat file to demonstrate the functionality of the SOM-based music content representation. The following lines of MATLAB code visualize the dataset stored in the musicLargeData.mat file using the SOM scheme:

```
load(['..' filesep 'data' filesep 'musicLargeData'])
musicVisualizationDemoSOM(FeaturesDir, FileNames, 8, 0)
```

[2] We have selected to use LDA as a preprocessing step. The LDA method is first employed to reduce the dimensionality of the feature space to 10 dimensions and the SOM is then trained on a 10-dimensional feature space. In this way, the training time of the SOM is reduced significantly.

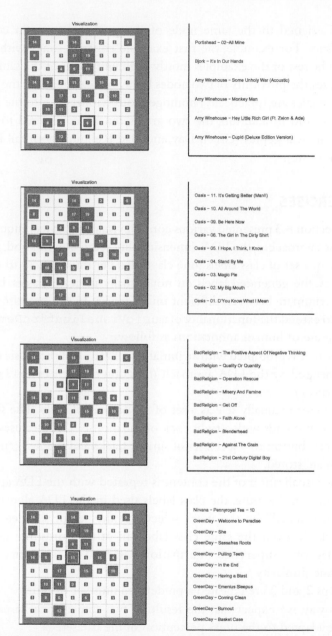

Figure 8.5 Visualization of selected nodes of the SOM of the data in the `musicLarge Data.mat` dataset.

The grid size for this example is 8×8. Figure 8.5 presents some node visualization examples. It can be seen that in some cases, the music tracks that were assigned to the same node of the grid share certain common characteristics. For example, the first example contains songs with female vocalists. The rest of the examples mostly highlight songs of the same artist. In some cases, the proximity of two nodes has a physical meaning: the 3rd and the 4th examples are (practically) homogeneous with respect to the identity of the artist, and in addition, the two artists can be considered to belong to similar musical genres (Green Day and Bad Religion both fall into the punk-rock genre).

8.4. EXERCISES

1. (D5) Section 8.3 presented various content visualization techniques. One of those approaches, the LDA dimensionality reduction method, requires the use of a set of class labels (one class label per music track) to operate. In a way, the generated labels do not actually indicate classes, but simply discriminate between different music tracks. In this exercise you are asked to extend the functionality of musicVisualizationDemo(), by making use of human annotations as follows:
   - The user is prompted with an initial visualization of the music content, generated by the LDA approach (function musicVisualization Demo()).
   - The user manually selects a set of *similar musical tracks*. The selection starts and ends with a right click of the mouse. The user uses the left mouse button to select a set of similar tracks and the respective track IDs are stored.
   - The visualization of the content is repeated with the LDA approach. However, this time, the class labels used in the LDA algorithm are "enchanched" by the user's selections: the set of selected musical tracks is given the same class label. In other words, the user provides some supervised information based on his/her perception of music similarity.
   - Steps 2 and 3 are repeated a predefined number of times.
   In this way, we expect that the resulting two-dimensional visualization will be adapted to the user's perception of the dataset.
2. (D3) Function musicVisualizationDemo() computes a projection matrix for each one of the adopted linear dimensionality reduction approaches (random projection, PCA, and LDA). As described in Section 8.3, for the first two methods (random projection and PCA) the long-term averages of the feature statistics of each music track are used,

whereas LDA uses the mid-term feature vectors of each track. Change the code of `musicVisualizationDemo()` so that random projection and PCA also use the mid-term feature vectors, instead of their averages. Can you see any differences in the resulting visualization results? Justify your answer. Use the `musicSmallData.mat` training. Note: In all cases, the final representation is based on the projection of long-term averages; what we ask here is to compute the projection transform based on the mid-term features.

3. (D4) Multidimensional scaling (MDS) is yet another widely used technique for data visualization, which has also been used in music information retrieval applications [139]. It focuses on representing the (dis) similarities amongst samples of a dataset, by means of a distance matrix. In MATLAB, MDS is implemented in the `mdscale()` function of the Statistics Toolbox. Apply `mdscale()` on the `musicSmallData.mat` dataset to visualize the respective music tracks in the 2D space. For more information on the `mdscale()` function, you may need to read the respective online MATLAB documentation.[3] *Hint 1: You first need to perform feature normalization on the initial dataset (zero mean and unity standard deviation). Hint 2: Use* `pdist()` *to compute a distance matrix between the feature vectors of the dataset.*

4. (D5) The Isomap method has also been used to visualize music content [120]. The Isomap algorithm [145] is actually a non-linear extension of the multidimensional scaling (MDS) method. Use the MATLAB implementation of the Isomap algorithm in http://isomap.stanford.edu/ to visualize the dataset stored in the `musicSmallData.mat` file. Note that the `Isomap()` function expects a distance matrix as an input argument (you will need to carefully read the documentation before you proceed). *Hint: You first need to perform feature normalization (to zero mean and unity standard deviation) on the initial dataset.*

5. (D3) In this exercise the reader is encouraged to experiment with all visualization approaches described in Section 8.3, using their own music data. To this end, you must first perform feature extraction on a directory of music data using the `featureExtractionDir()` function.

6. (D4) The SOM-based visualization approach (Section 8.3.3) leads to a discretized representation of the final 2D space, which means that usually more than one sample (in our case, music tracks) will be mapped to a single SOM node. Combine the SOM and LDA approaches for the purposes of

---

[3] http://www.mathworks.com/help/stats/mdscale.html.

music content visualization. The core of the new visualization function must be the SOM approach. However, when the user selects a particular SOM grid node, instead of simply printing the respective song titles (as happens in `musicVisualizationDemoSOM()`), your function must use the LDA dimensionality reduction approach to plot the respective song titles in the 2D space. Obviously, the LDA algorithm must be trained on the samples that are mapped to the selected SOM grid (and not on the whole dataset).

7. (D5) Implement a query-by-humming system that broadly speaking, follows the guidelines of the MIREX 2011 Query-by-Singing/Humming contest http://www.music-ir.org/mirex/wiki/2011:Query-by-Singing/Humming_Results#Task_Descriptions. The following steps outline the tasks that you need to complete:

   (a) Download Roger Jang's MIR-QBSH corpus from http://mirlab.org/dataSet/public/MIR-QBSH-corpus.rar. This consists of the sung tunes of 48 tracks. The sung melodies are organized in folders based on the year of recording and the singer. The name of each WAVE file reveals the melody to which it refers. Each audio file is accompanied by an extracted pitch sequence.

   (b) Place all audio and pitch files in a single folder. To avoid duplicate file names, use a script to rename each file so that the recording year and the person's id become part of the string of the file name.

   (c) Match each pitch sequence against all others using the Itakura local path constraints. Sort the matching costs in ascending order and examine the best 10 results. If the majority of results stem from the same melody, a success is recorded.

   (d) Repeat the previous step for all pitch sequences and compute the accuracy of the system.

   In a way, we are using a version of the leave-one-out method for performance validation. Repeat with the Sakoe local-path constraints and the Smith-Waterman algorithm. For the latter, you will need to devise a modified similarity function (why?). Finally, experiment with endpoint constraints (see also the related exercises in Chapter 7).

8. (D5) Develop a MATLAB function that converts a WAVE file to a sequence of notes. It is assumed that we are dealing with monophonic recordings of a single instrument. Your function will need to perform the following tasks:

   (a) Load the WAVE file.

   (b) Perform pitch tracking (see also Chapter 4).

(c) Quantize each value of the pitch sequence to the closest semitone of the chromatic scale.

(d) Convert the sequence of quantized pitch values to a sequence of vectors of the form $[p\ t]^T$, where p is the pitch value and t is its duration (you will need to merge successive pitch values to a single vector if they are equal).

(e) Drop short notes. Any note with a duration shorter than a user-defined threshold is eliminated.

(f) The remaining sequence of vectors is stored in a text file.

In addition, develop a second function that loads the previously generated text file, converts it to the MIDI format, and uses an external MIDI player to playback the result. You may need to download a 3d-party, publicly available MATLAB toolbox to implement the MIDI functionality. Alternatively, you can reproduce the stored file as a sequence of simple tones. The result might not please the ear but it will give you an idea of how successful your music transcription function has been.

The Matlab Audio Analysis Library

This book is accompanied by a MATLAB software library, to assist with the reproducibility of the methods presented in the book and as a toolbox for readers wishing to embark on their own projects. Each function in the library contains a description of its functionality. In this appendix, we present a complete list of the m-files and their descriptions. After you download the software library from the companion site, all the available m-files will be decompressed in the library subfolder of the software distribution (see Table A.1).

Table A.1 List of All Functions Included in the MATLAB Audio Analysis Library Provided with the Book

Name	Description	Chapter
classifyKNN_D_Multi()	Classifies an unknown sample using the k-NN algorithm, in its multi-class mode. Returns probability estimates.	5, 6
computePerformance-Measures()	Computes the confusion matrix and performance measures of a classification process.	5
dctCompress() and dctDecompress	Demonstrate the use of the DCT for compressing and decompressing an audio signal.	3
dynamicTimeWarpingSakoe-Chiba()	Computes the Dynamic Time Warping cost between two feature sequences based on the Sakoe-Chiba local path constraints.	7

(*Continued*)

Table A.1 Continued

Name	Description	Chapter
dynamicTimeWarping- Itakura()	Computes the Dynamic Time Warping cost between two feature sequences based on the Itakura local path constraints.	7
evaluateClassifier()	Implements the repeated hold out and leave-one-out validation methods.	5
feature_chroma_vector()	Computes the chroma vector of a short-term window.	4
feature_energy()	Computes the energy of a short-term window.	4
feature_energy_entropy()	Computes the entropy of energy of a short-term window.	4
featureExtractionDir()	Extracts mid-term features for a list of WAVE files stored in a given folder.	8
featureExtractionFile()	Reads a WAVE file and computes audio feature statistics on a mid-term basis.	4, 5, 6
feature_harmonic()	Computes the harmonic ratio and fundamental frequency of a window (autocorrelation method).	4
feature_mfccs_init()	Initializes the computation of the MFCCs (see also feature_mfccs()).	4
feature_mfccs()	Computes the MFCCs of a short-term window[a].	4
feature_spectral_centroid()	Computes the spectral centroid of a short-term window.	4
feature_spectral_entropy()	Computes the spectral entropy of a short-term window.	4
feature_spectral_flux()	Computes the spectral flux of a short-term window.	4
feature_spectral_rolloff()	Computes the spectral rolloff of a short-term window.	4
feature_zcr()	Computes the zero-crossing rate of a short-term window.	4

(*Continued*)

Table A.1 Continued

Name	Description	Chapter
fftExample()	Demonstrates how to use the getDFT() function.	3
fileClassification()	Demonstrates the classification of a WAVE file as a whole (not to be confused with mtFileClassification() which performs segmentation and classification).	5
fld()	Finds a linear discriminant subspace using the LDA algorithm. Used for dimensionality reduction in the context of music visualization. *This m-file has not been implemented by the authors, but was taken from*[b].	8
getDFT()	Returns the (normalized) magnitude of the DFT of a signal.	3, 4
k-NN_model_add_ class()	Adds an audio class to a k-NN classification setup. As the k-NN classifier requires no actual training, the only function it performs is a feature extraction stage for a set of WAVE files, stored in a given directory.	5
kNN_model_load()	Loads a k-NN classification setup, i.e. a feature matrix for each class, along with the respective normalization parameters (means and standard deviations of the features).	5
mp3toWav()	Performs MP3 to WAVE conversion with the FFmpeg command-line tool.	2
mp3toWavDIR()	Transcodes each MP3 of a given folder to the WAVE format, using the FFmpeg command-line tool.	2

(Continued)

Table A.1 Continued

Name	Description	Chapter
mtFeature-Extraction()	Computes the mid-term statistics for a set of sequences of short-term features. It returns a matrix, whose columns contain the vectors of mid-term feature statistics.	4
mtFile-Classification()	Splits an audio signal into fixed-size segments and classifies each segment separately (fixed-size window segmentation).	5, 6
musicMeterTempo-Induction()	Performs joint estimation of the music meter and tempo of a music recording.	8
musicThumbnailing()	Extracts pairs of thumbnails from music recordings.	8
musicVisualization-Demo()	Demonstrates three linear dimensionality reduction methods for music content visualization (random projection, PCA, and LDA).	8
musicVisualization-DemoSOM()	Demonstrates SOM-based music content visualization.	8
plotFeaturesFile()	Plots a given feature sequence that has been computed over a WAVE file.	4
printPerformance-Measures()	Prints a table of classification performance measures (confusion matrix, recall, etc.) in $\LaTeX$ format.	5
readWavFile()	Demonstrates how to read the contents of a WAVE file, using two different modes: (a) all the contents of the WAVE file are loaded, and (b) blocks of data are read and each block is processed separately.	2
readWavFileScript()	Generates experiments that measure the elapsed time of different WAVE file I/O approaches.	2
scaledBaumWelch-ContObs()	Implements the scaled version of the *Baum-Welch* algorithm (continuous features).	7

(Continued)

Table A.1 Continued

Name	Description	Chapter
scaledBaumWelchDisObs()	Implements the scaled version of the *Baum-Welch* algorithm (discrete observations).	7
scaledViterbiContObs()	Implements the Viterbi algorithm for continuous features.	7
scaledViterbiDisObs()	Implements the Viterbi algorithm for discrete observations.	7
scriptClassification-Performance()	Loads a k-NN classification setup (stored in a mat file) and extracts the respective classification performance measures. For the best value of k, it prints the respective confusion matrix and class-specific performance measures.	5
segmentationCompare-Results()	Visualizes two different segmentation results for the sake of comparison.	6
segmentationPlot-Results()	Provides a simple user interface to view and listen to the results of a segmentation-classification procedure.	6
segmentationProbSeq()	Segments an audio stream based on the estimated posterior probabilities for each class. Implements: (a) naive merging, and (b) viterbi-based probability smoothing. To be called after mtFileClassification().	6
segmentation-SignalChange()	Basic unsupervised signal change segmentation (no classifier needed).	6
showHistogram-Features()	This auxiliary function is used to plot the histograms of a particular feature for different audio classes. It has been used to generate the histograms of Chapter 4.	4

(Continued)

Table A.1 Continued

Name	Description	Chapter
silenceDetector-Utterance()	Computes the endpoints of a *single* speech utterance. Based on Rabiner and Schafer, Theory and Applications of Digital Speech Processing, Section 10.3.	6
silenceRemoval()	Applies a semi-supervised algorithm for detecting speech segments (removing silence) in an audio stream stored in a WAVE file.	6
smithWaterman()	Implements the Smith–Waterman algorithm for sequence alignment.	7
soundOS()	An alternative to the MATLAB sound() function, in case problems are encountered in Linux-based systems.	2
speakerDiarization()	Implements a simple unsupervised speaker diarization procedure.	6
stFeature-Extraction()	Breaks an audio signal into possibly overlapping short-term windows and computes sequences of audio features. It returns a matrix whose rows correspond to the extracted feature sequences.	4
stpFile()	Demonstrates the short-term processing stage of an audio signal.	2
viterbiBestPath()	Finds the most-likely state sequence given a matrix of probability estimations. Used for *smoothing* segmentation results.	6
viterbiTrainingDo()	Implements the Viterbi training scheme in the case of discrete observations.	7

(*Continued*)

Table A.1 Continued

Name	Description	Chapter
viterbiTrainingMultiCo()	Implements the Viterbi training scheme in the case of continuous, multidimensional features, under the assumption that the density function at each state is Gaussian.	7
viterbiTraining-MultiCoMix()	Implements the Viterbi training scheme in the case of Gaussian mixtures.	7

[a]Partly based on Slaney's Auditory Toolbox [29].

[b]Mathworks File Exchange, Fisher Linear Discriminant Analysis, by Sergios Petridis: http://www.mathworks.com/matlabcentral/fileexchange/38950-fischer-linear-dicriminant-analysis

Apart from the provided m–files in the accompanying library, there are also some data files that either contain sample audio data or audio features stored in mat files. The most important of these data files are shown in the following table:

Table A.2 List of Data Files that are Available in the Library that Accompanies the Book. Some of the Files (Especially Those Related to Training Features used by *k*-NN Classifiers) are Stored in the library Folder, while most of them are Stored in the data Folder of our Software Distribution

Name	Description
data/1WORD.WAV, data/3WORDS.WAV	Speech examples that can be used in silence detection or speech filtering (Chapter 6).
data/4ClassStream.wav, data/4ClassStreamGT.mat	4–class (female speech, male speech, silence, and music) example to be used for supervised segmentation methods (Chapter 6). The mat file contains the respective ground truth.
data/clarinet *.wav	Clarinet sounds.
data/diarization Example.wav	Audio example for speaker diarization (Chapter 6).

(Continued)

Table A.2 Continued

Name	Description
data/WindInstrument-Pitch.mat	A collection of pitch-tracking sequences from clarinet sounds (Chapter 7.)
data/KingGeorge-Speech_1939_53sec.wav, data/KingGeorge-Speech_1939_small.wav, data/DubaiAirport.wav	Three general-purpose speech files (used for silence detection, segmentation, filtering, and so on).
data/musicLargeData.mat, data/musicSmallData.mat	Two datasets of mid-term features extracted from 300 and 40 music tracks, respectively. Used for music visualization tasks (Chapter 8).
data/speech_music_sample.wav	An audio stream of speech and music segments. Used for speech-music segmentation methods (Chapter 6).
data/topGear.wav, data/topGear.mat	An audio stream from a TV show with respective ground truth. Used by signal change detection methods (Chapter 6).
library/model*.mat	All these mat files contain the training data (feature vectors) for the respective classification tasks. A complete list of these mat files is presented in Table 5.1.

Audio-Related Libraries and Software

In this appendix we present a number of audio analysis libraries and methodologies for MATLAB and other programming languages. In addition, we present related (non-audio) libraries and packages that could be used in the context of intelligent signal analysis, e.g. numerical analysis, general signal processing, multimedia file I/O, pattern recognition, and data mining resources. Although we have primarily focused on MATLAB-related libraries, we also give an idea of relevant resources in Python and C/C++.

For the MATLAB environment, we present separately the audio-specific and general pattern recognition libraries, while Python and C++ are presented in a single list. We have chosen to include Python-related approaches because of the language's similarities to MATLAB and because of its wide acceptance in the scientific community. In addition, C/C++ will always be a lower level (and less easy-to-handle) solution for signal processing, which, however, leads to faster programs compared to MATLAB and Python.

B.1. MATLAB

Table B.1 presents a list of MATLAB libraries on audio and speech analysis that are available on the Web. Note that two of these libraries are focused on music information retrieval, one is speech-oriented and only one covers generic audio analysis.

Furthermore, in this book we presented a number of *pattern recognition* methods that reside in MATLAB toolboxes, e.g. support vector machines, decision trees, etc. However, there are also plenty of related libraries on the Internet Web. Table B.2 we present a short list of MATLAB libraries related to pattern recognition and machine learning techniques.

Table B.1 MATLAB Libraries—Audio and Speech

Name	Description
Auditory Toolbox, Version 2, by Malcolm Slaney	A MATLAB library that focuses on representing stages of the human auditory analysis system, https://engineering.purdue.edu/malcolm/interval/1998-010/
MIRtoolbox	A MATLAB library that deals with the extraction of musical features such as tonality, rhythm, etc. https://www.jyu.fi/hum/laitokset/musiikki/en/research/coe/materials/mirtoolbox
VOICEBOX	A speech processing toolbox for MATLAB, http://www.ee.ic.ac.uk/hp/staff/dmb/voicebox/voicebox.html
MA Toolbox	A Matlab toolbox for music analysis, http://www.pampalk.at/ma/

Table B.2 MATLAB Libraries—Pattern Recognition and Machine Learning

Name	Description
Pattern Recognition Toolbox (PRT) for MATLAB	Wide range of pattern recognition techniques (classification, clustering, outlier removal, regression, etc.). MIT License. Parts of the PRT may require a C/C++ compiler to be installed, http://www.newfolderconsulting.com/prt/download
LibSVM	An integrated software for support vector classification, regression, and pdf estimation. It supports multi-class classification. It is not written in MATLAB. However, it supports APIs for other languages and packages (MATLAB, Python, Weka, R, Octave, Java, etc.), [146], http://www.csie.ntu.edu.tw/~cjlin/libsvm/
Hidden Markov Model (HMM) Toolbox for Matlab, MIT	Supports inference and learning for HMMs that emit discrete observations or continuous observations based on Gaussian pdfs and mixtures of Gaussian pdfs, http://www.cs.ubc.ca/~murphyk/Software/HMM/hmm.html
Pattern Recognition with Matlab—Online companion material of the book [40]	Consists of a set of functions that cover various stages of the design of a pattern recognition system.

removelastskip

B.2. PYTHON

Python is a high-level programming language that has been attracting increasing interest in the scientific community during the past few years. A wide range of available packages can be used in order to develop a MATLAB-like functionality in Python. The main reason that Python is an attractive programming language for signal processing applications is that it is characterized by a balance of high-level and low-level programming features. In particular, one can easily write algorithms with less lines of code (compared to C/C++), while at the same time, the initial problem of low speed is partly solved by the application of optimization procedures on higher level objects. For example, it is possible to speed up the execution of Python code by vectorizing the respective algorithm. Another great advantage of Python is that there exists an impressive number of libraries that provide functionalities related to scientific programming.

Unlike MATLAB, Python is non-commercial but emphasizes on *portability*: On the other hand, MATLAB provides Matlab Component Runtime, which can be a more complicated solution compared to Python's cross-platform nature. On the other hand, MATLAB, compared to Python, is easier to learn (especially for non-programmers), it provides an easy-to-use integrated development environment (IDE) and easier plotting functionalities. In addition, MATLAB is common in almost every scientific field. Therefore, there exist plenty of available MATLAB-compatible software resources that address the needs of a large scientific community.

Table B.3 presents a list of Python packages and libraries that can be used for the development of audio analysis applications.

Ubuntu users will find most of the Python packages in the Ubuntu repository. For example, NumPy can be installed by simply typing the following command in a terminal: `apt-get install python-numpy`.

Table B.3 A List of Python Packages and Libraries that can be Used for Audio Analysis and Pattern Recognition Applications

Name	Description
NumPy	A library for scientific computing http://www.numpy.org/, with convenient and efficient N-dimensional array manipulation.
SciPy	A library for mathematics, science, and engineering http://www.scipy.org/, based on NumPy. Also used for the I/O of WAVE files.
MLPy	A library for machine learning, built on top of NumPy and SciPy. http://mlpy.sourceforge.net/
matplotlib	A library for 2-D plotting in Python. http://matplotlib.org/
ALSAaudio	A library of wrappers for accessing the ALSA API from Python and necessary for handling audio input/output. http://pyalsaaudio.sourceforge.net/
Yaafe	A Python library for audio feature extraction. http://yaafe.sourceforge.net/

B.3. C/C++

C and C++ are widely acknowledged programming languages used for computationally demanding signal processing applications. For real-time signal processing applications that analyze voluminous data, C/C++ is usually the most suitable solution. Of course, both languages demand for more experienced programmers and additional development costs, compared to the Python or MATLAB approaches. However, it is beyond doubt that C/C++ have been widely used in several signal processing and machine learning applications during the last 20 years. In Table B.4 we present a list of representative C/C++ packages and libraries that can be used for building audio analysis applications.

> *Note:* Apart from the general pattern recognition and machine learning libraries listed in the aforementioned tables, a good resource for open-source implementations of non-trivial machine learning algorithms, toolboxes, etc. is given by the Journal of Machine Learning Research in http://jmlr.org/mloss/. This resource covers a wide range of programming languages (C++, Python, Java, etc.) and machine learning fields (general pattern recognition, classification, regression, clustering, large-scale data processing, etc.).

Table B.4 Representative Audio Analysis and Pattern Recognition Libraries and Packages Written in C++

Name	Description
CLAM (C++ Library for Audio and Music)	A framework for research/development in the audio and music domain. Provides the means to perform complex audio signal analysis, transformations, and synthesis. Can be used as a library and a graphical tool. http://clam-project.org/
MARSYAS (Music Analysis, Retrieval, and Synthesis for Audio Signals)	An open-source library for audio analysis, mostly focused on Music Information Retrieval [147,148]. In existence since 1998, it has been used for a variety of academic and industrial projects. Written in C++, but also ported to Java. Can also be installed with Python bindings. http://marsyas.info/
aubio	A tool written in C for basic audio analysis: pitch tracking, onset detection, extraction of MFCCs, beat and meter tracking, etc. Provides wrappers for Python. http://aubio.org/
Maaate	A C++ toolkit to parse and analyze audio data in the compressed/frequency domain. http://maaate.sourceforge.net/
Synthesis ToolKit in C++ (STK)	Audio signal processing and algorithmic synthesis methods written in C++. Focuses on music synthesis functionality. https://ccrma.stanford.edu/software/stk/
Vamp plugin system	A set of audio analysis plugins. Provides a C++ API but also a Python wrapper and an interface that permits Java applications to run native Vamp plugins. For example, a Vamp plugin can be used inside an audio editor (e.g. Audacity) to enhance the visualization of information. http://www.vamp-plugins.org/
dlib	A cross-platform C++ library that covers a wide range of machine learning, image processing, linear algebra, and general-purpose algorithms [149]. Licensed under the Boost Software License. http://dlib.net/

APPENDIX *C*

Audio Datasets

Several datasets and benchmarks that focus on audio analysis tasks are available on the Web. The diversity of the datasets is high with respect to: size, level of annotation, and addressed audio analysis tasks. For example, there are datasets for general audio event classification and segmentation; musical genre classification; speech emotion recognition; speech vs music discrimination; speaker diarization; speaker identification, etc. In addition, these datasets may or may not contain other non-audio media types (e.g. textual or visual information). It is hard to provide a complete list of all available datasets related to audio analysis. Table C.1 simply presents some representative datasets, which are available on the Web, for a selected audio analysis tasks.

Table C.1 A Short List of Available Datasets for Selected Audio Analysis Tasks

Name	Task	Description
GTZAN Genre Collection	Musical genre classification	Consists of 1000 audio tracks (30 s each). Contains 10 genres (100 tracks each)[a].
GTZAN Music Speech Collection	Speech-music discrimination	Consists of 120 tracks (30 s each). Each class has 60 files[b].
Magnatagatune	Several MIR tasks	Covers a wide range of MIR annotations: artist identification, mood classification, instrument identification, music similarity, etc[c].
Free Music Archive	Musical genre classification	An interactive library of high-quality, legal audio downloads. It is a good resource for music data. Organized in genres[d].

(Continued)

Table C.1 Continued

Name	Task	Description
MIREX (Music Information Retrieval Evaluation eXchange)	Several MIR tasks	A community-based formal framework for the evaluation of a wide range of techniques in the domains of Music Information Retrieval and Digital Libraries of Music. Covers a wide range of tasks, including: cover song identification, onset detection, symbolic melodic similarity, chord estimation, beat tracking, tempo estimation, genre classification, tag classification, etc. A MIREX contest is organized annually as a satellite event of the International Society for Music Information Retrieval (ISMIR) Conference[e].
Million Song Dataset [150]	Several MIR tasks	A collection of audio features and metadata for a million contemporary popular music tracks. Does not include audio, only features. Can be used for several MIR tasks: segmentation, tagging, year recognition, artist recognition, cover song recognition, etc[f].
Canal 9 Political Debates	Speaker diarization	A collection of 72 political debates recorded by the Canal 9 local TV and radio station in Valais, Switzerland. Audio-visual recordings. Three to five speakers in each recording. 42 h duration in total[g].
NIST Speaker Recognition Evaluation (SRE)	Speaker recognition	NIST (National Institute of Standards and Technology of the US Department of Commerce) has been coordinating speaker recognition evaluations since 1996. The evaluation culminates with a follow-up workshop, where NIST reports the official results and researchers share their findings[h].
PRISM (Promoting Robustness in Speaker Modeling)	Speaker recognition	A dataset for speaker recognition based on NIST, enhanced with new denoising and dereverberation tasks. Includes signal variation already seen in one or more NIST SREs, namely: language, channel type, speech style, and vocal effort level[i].

(Continued)

Table C.1 Continued

Name	Task	Description
MediaEval Benchmark	Several multimedia analysis tasks	This benchmarking initiative has been taking place since 2010 and focuses on several tasks that require the analysis of image, text, and audio. Some of the tasks that include audio information (among other types of media) are: geo-coordinate prediction for social multimedia, violence detection in movies, spoken web search, soundtrack selection for commercials, etc[j].
The ICML 2013 Whale Challenge— Right Whale Redux	Audio classification	Dataset built in the context of whales sound classification for big data mining. Several similar datasets of sea mammal sounds have also been created in the past[k].
Berlin Database of Emotional Speech	Speech emotion recognition	A German database of acted emotional speech. Seven emotional states: neutral, anger, fear, joy, disgust, boredom, and sadness. Ten actors[l].

[a] http://marsyas.info/download/data_sets/.

[b] http://marsyas.info/download/data_sets/.

[c] http://musicmachinery.com/2009/04/01/magnatagatune-a-new-research-data-set-for-mir/.

[d] http://freemusicarchive.org/.

[e] http://www.music-ir.org/mirex/wiki/MIREX_HOME.

[f] http://labrosa.ee.columbia.edu/millionsong/.

[g] http://www.idiap.ch/scientific-research/resources/canal-9-political-debates.

[h] http://www.nist.gov/itl/iad/mig/sre.cfm.

[i] http://code.google.com/p/prism-set/.

[j] http://www.multimediaeval.org/.

[k] http://www.kaggle.com/c/the-icml-2013-whale-challenge-right-whale-redux.

[l] http://www.expressive-speech.net/.

Notes:

- Speech emotion recognition has gained significant research interest during the last decade. Therefore, there are several databases, not always based only on speech but on visual cues. It is beyond the purpose of this book to provide a complete report on these datasets. However, a rather detailed description of the available audio-visual emotional databases can be found in http://emotion-research.net/wiki/Databases.
- The reader may easily conclude that we have not mentioned databases that focus on Automatic Speech Recognition (ASR). This is because the purpose of the book is to focus on general audio analysis tasks and not on the transcription of spoken words.

Bibliography

[1] John G. Proakis, Dimitris K. Manolakis, Digital Signal Processing, fourth ed., Pearson Education, 2009.

[2] Tim Kientzle, A Programmer's Guide to Sound with Cdrom, Addison-Wesley Longman Publishing Co. Inc., 1997.

[3] Karlheinz Brandenburg, Mp3 and aac explained, in: Audio Engineering Society Conference: 17th International Conference: High-Quality Audio Coding, 1999.

[4] Ted Painter, Andreas Spanias, Perceptual coding of digital audio, Proceedings of the IEEE 88 (4) (2000) 451–515.

[5] S. Theodoridis, K. Koutroumbas, Pattern Recognition, fourth ed., Academic Press, Inc., 2008.

[6] M. Frigo, S.G. Johnson, Fftw: an adaptive software architecture for the fft, in: Proceedings of the International Conference on Acoustics, Speech, and, Signal Processing, 1998, pp. 1381–1384.

[7] Kim Hyoung-Gook, Moreau Nicolas, Thomas Sikora, MPEG-7 Audio and Beyond: Audio Content Indexing and Retrieval, John Wiley & Sons, 2005.

[8] Lawrence R. Rabiner, Ronald W. Schafer, Introduction to digital speech processing, Foundations and Trends in Signal Processing, Now Publishers Inc, 2007.

[9] Emmanuel C. Ifeachor, Barrie W. Jervis, Digital Signal Processing: A Practical Approach, Pearson Education, 2002.

[10] Alan V. Oppenheim, Ronald W. Schafer, John R. Buck, et al., Discrete-Time Signal Processing, vol. 5, Prentice Hall, Upper Saddle River, 1999.

[11] Nasir Ahmed, T. Natarajan, Kamisetty R. Rao, Discrete cosine transform, IEEE Transactions on Computers 100 (1) (1974) 90–93.

[12] Miroslav D. Lutovac, Dejan V. Tošić, Brian Lawrence Evans, Filter design for signal processing using MATLAB and Mathematica, Prentice Hall, 2001.

[13] Edward Kamen, Bonnie Heck, Fundamentals of Signals and Systems: With MATLAB Examples, Prentice Hall PTR, 2000.

[14] G. Tzanetakis, P. Cook, Musical genre classification of audio signals, IEEE Transactions on Speech and Audio Processing 10 (5) (2002) 293–302.

[15] George Tzanetakis, Gtzan genre collection. <http://marsyas.info/download/data_sets/>.

[16] C. Panagiotakis, G. Tziritas, A speech/music discriminator based on rms and zero-crossings, IEEE Transactions on Multimedia 7 (1) (2005) 155–166.

[17] Lee Daniel Erman, An environment and system for machine understanding of connected speech (Ph.D thesis), Stanford, CA, USA, 1974 (AAI7427012).

[18] E. Scheirer, M. Slaney, Construction and evaluation of a robust multifeature speech/music discriminator, in: ICASSP '97: Proceedings of the 1997 IEEE International Conference on Acoustics, Speech, and Signal Processing, IEEE Computer Society, Washington, DC, USA, 1997, pp. 1331.

[19] Lawrence R. Rabiner, Marvin R. Sambur, An algorithm for determining the endpoints of isolated utterances, Bell System Technical Journal 54 (2) (1975) 297–315.

[20] T. Giannakopoulos, A. Pikrakis, S. Theodoridis, Gunshot detection in audio streams from movies by means of dynamic programming and bayesian networks, in: 33rd International Conference on Acoustics, Speech, and Signal Processing, ICASSP08, 2008.

[21] Theodoros Giannakopoulos, Aggelos Pikrakis, Sergios Theodoridis, A multi-class audio classification method with respect to violent content in movies, using bayesian networks, in: IEEE International Workshop on Multimedia, Signal Processing, MMSP07, 2007.

[22] Hemant Misra, Shajith Ikbal, Hervé Bourlard, Hynek Hermansky, Spectral entropy based feature for robust ASR, in: Proceedings of the 2004 IEEE International Conference on Acoustics, Speech, and Signal Processing, ICASSP'04, vol. 1, IEEE, 2004, pp. I–193.

[23] A. Pikrakis, T. Giannakopoulos, S. Theodoridis, A computationally efficient speech/music discriminator for radio recordings, in: International Conference on Music Information Retrieval and Related Activities, ISMIR06, 2006.

[24] A. Pikrakis, T. Giannakopoulos, S. Theodoridis, A speech/music discriminator of radio recordings based on dynamic programming and bayesian networks, IEEE Transactions on Multimedia 10 (5) (2008) 846–857.

[25] Steven Davis, Paul Mermelstein, Comparison of parametric representations for monosyllabic word recognition in continuously spoken sentences, IEEE Transactions on Acoustics, Speech and Signal Processing 28 (4) (1980) 357–366.

[26] Leo L. Beranek, Acoustical Measurements, revised ed., American Institute of Physics, Cambridge, MA, 1988.

[27] David Pearce, Hans günter Hirsch, The aurora experimental framework for the performance evaluation of speech recognition systems under noisy conditions, in: ISCA ITRW ASR2000, 2000, pp. 29–32, Ericsson Eurolab Deutschland Gmbh.

[28] Theodoros Giannakopoulos, Sergios Petridis, Unsupervised speaker clustering in a linear discriminant subspace, in: Proceedings of the 2010 Ninth International Conference on Machine Learning and Applications, ICMLA '10, 2010, pp. 1005–1009.

[29] M. Slaney, Auditory Toolbox, Version 2. Technical Report, Interval Research Corporation, 1998.

[30] Gregory H. Wakefield, Mathematical representation of joint time-chroma distributions, in: SPIE's International Symposium on Optical Science, Engineering, and Instrumentation, International Society for Optics and Photonics, 1999, pp. 637–645.

[31] Mark A. Bartsch, Gregory H. Wakefield, Audio thumbnailing of popular music using chroma-based representations, IEEE Transactions on Multimedia 7 (1) (2005) 96–104.

[32] Mark A. Bartsch, Gregory H. Wakefield, To catch a chorus: using chroma-based representations for audio thumbnailing, in: 2001 IEEE Workshop on the Applications of Signal Processing to Audio and Acoustics, IEEE, 2001, pp. 15–18.

[33] Meinard Müller, Frank Kurth, Michael Clausen, Audio matching via chroma-based statistical features, in: Proceedings of ISMIR, London, GB, 2005, pp. 288–295.

[34] Hyoung-Gook Kim, Nicolas Moreau, Thomas Sikora, MPEG-7 Audio and Beyond: Audio Content Indexing and Retrieval, Wiley, 2006.

[35] Manfred R. Schroeder, Period histogram and product spectrum: new methods for fundamental-frequency measurement, Journal of the Acoustical Society of America 43 (1968) 829.

[36] William B. Kendall, A new algorithm for computing correlations, IEEE Transactions on Computers 23 (1) (1974) 88–90.

[37] Tero Tolonen, Matti Karjalainen, A computationally efficient multipitch analysis model, IEEE Transactions on Speech and Audio Processing 8 (6) (2000) 708–716.

[38] David G. Stork, Richard O. Duda, Peter E. Harr, Pattern Classification, second ed., Wiley-Interscience, 2000.

[39] Christopher M. Bishop, Pattern Recognition and Machine Learning (Information Science and Statistics), Springer-Verlag, New York, Inc., 2006.

[40] A. Pikrakis, S. Theodoridis, K. Koutroumbas, D. Kavouras, Introduction to Pattern Recognition: A Matlab Approach, Academic Press, 2009.

[41] Irina Rish, An empirical study of the naive bayes classifier, in: IJCAI 2001 Workshop on Empirical Methods in Artificial Intelligence, vol. 3, 2001, pp. 41–46.

[42] Andrew McCallum, Kamal Nigam, et al., A comparison of event models for naive bayes text classification, in: AAAI-98 Workshop on Learning for Text Categorization, vol. 752, Citeseer, 1998, pp. 41–48.

[43] John S. Garofolo et al., Timit acoustic-phonetic continuous speech corpus. <http://www.ldc.upenn.edu/Catalog/CatalogEntry.jsp?catalogId=LDC93S1>.

[44] B.R. Kowalski, C.F. Bender, K-nearest neighbor classification rule (pattern recognition) applied to nuclear magnetic resonance spectral interpretation, Analytical Chemistry 44 (8) (1972) 1405–1411.

[45] J. Fürnkranz, Round robin classification, Journal of Machine Learning Research 2 (2002) 721–747.

[46] R. Rifkin, A. Klautau, In defense of one-vs-all classification, Journal of Machine Learning Research 5 (2004) 101–141.

[47] G. Tsoumakas, I. Katakis, Multi-label classification: an overview, International Journal of Data Warehousing and Mining (IJDWM) 3 (3) (2007) 1–13.

[48] Kazuo Hattori, Masahito Takahashi, A new edited k-nearest neighbor rule in the pattern classification problem, Pattern Recognition 33 (3) (2000) 521–528.

[49] Hanan Samet, K-nearest neighbor finding using maxnearestdist, IEEE Transactions on Pattern Analysis and Machine Intelligence 30 (2) (2008) 243–252.

[50] L. Breiman, J.H. Friedman, R.A. Olshen, Charles J. Stone, Classification and regression trees, Wadsworth International Group (1984).

[51] Lior Rokach, Oded Z. Maimon, Data Mining with Decision Trees: Theory and Applications, vol. 69, World Scientific Publishing Company Incorporated, 2008.

[52] Ian H. Witten, Eibe Frank, Data Mining: Practical Machine Learning Tools and Techniques, second ed., Morgan Kaufmann Publishers Inc., 2005.

[53] C. Cortes, V. Vapnik, Support-vector networks, Machine Learning 20 (3) (1995) 273–297.

[54] C.J.C. Burges, A tutorial on support vector machines for pattern recognition, Data Mining and Knowledge Discovery 2 (2) (1998) 121–167.

[55] C.W. Hsu, C.J. Lin, A comparison of methods for multiclass support vector machines, IEEE Transactions on Neural Networks 13 (2) (2002) 415–425.

[56] S. Cheong, S.H. Oh, S.Y. Lee, Support vector machines with binary tree architecture for multi-class classification, Neural Information Processing-Letters and Reviews 2 (3) (2004) 47–51.

[57] David W. Aha, Dennis Kibler, Marc K. Albert, Instance-based learning algorithms, Machine Learning 6 (1) (1991) 37–66.

[58] J. Saunders, Real-time discrimination of broadcast speech/music, in: ICASSP '96: Proceedings of the Acoustics, Speech, and Signal Processing, 1996, pp. 993–996.

[59] K. El-Maleh, M. Klein, G. Petrucci, P. Kabal, Speech/music discrimination for multimedia applications, in: ICASSP '00: Proceedings of the 2000 IEEE International Conference on Acoustics, Speech, and Signal Processing, IEEE Computer Society, Washington, DC, USA, 2000, pp. 2445–2448.

[60] J. Ajmera, I. McCowan, H. Bourlard, Speech/music segmentation using entropy and dynamism features in a HMM classification framework, Speech Communication 40 (3) (2003) 351–363.

[61] Xavier Anguera Miro, Simon Bozonnet, Nicholas Evans, Corinne Fredouille, Gerald Friedland, Oriol Vinyals, Speaker diarization: a review of recent research, IEEE Transactions on Audio, Speech, and Language Processing 20 (2) (2012) 356–370.

[62] Frédéric Bimbot, Jean-François Bonastre, Corinne Fredouille, Guillaume Gravier, Ivan Magrin-Chagnolleau, Sylvain Meignier, Teva Merlin, Javier Ortega-García, Dijana Petrovska-Delacrétaz, Douglas A. Reynolds, A tutorial on text-independent speaker verification, EURASIP Journal on Applied Signal Processing 2004 (2004) 430–451.

[63] Bob L. Sturm, An analysis of the GTZAN music genre dataset, in: Proceedings of the Second International ACM Workshop on Music information Retrieval with User-Centered and Multimodal Strategies, MIRUM '12, ACM, New York, NY, USA, 2012, pp. 7–12.

[64] T. Giannakopoulos, S. Petridis, Fisher linear semi-discriminant analysis for speaker diarization, IEEE Transactions on Audio, Speech, and Language Processing 20 (7) (2012) 1913–1922.

[65] J. Picone, Continuous speech recognition using hidden markov models, Accoustic, Speech, Signal Processing (ASSP) Magazine 7 (3) (1990) 26–41.

[66] L.R. Rabiner, A tutorial on hidden markov models and selected applications in speech recognition, Proceedings of the IEEE 77 (2) (1989) 257–286.

[67] M. Plumpe, A. Acero, H. Hon, X. Huang, HMM-based smoothing for concatenative speech synthesis, in: Proceedings of ICSLP, vol. 6, Citeseer, 1998, pp. 2751–2754.

[68] S.E. Tranter, D.A. Reynolds, An overview of automatic speaker diarization systems, IEEE Transactions on Audio, Speech, and Language Processing 14 (5) (2006) 1557–1565.

[69] Hari Sundaram, Shih-Fu Chang, Video scene segmentation using video and audio features, 2000 IEEE International Conference on Multimedia and Expo, ICME 2000, vol. 2, IEEE, 2000, pp. 1145–1148.

[70] Christopher Harte, Mark Sandler, Martin Gasser, Detecting harmonic change in musical audio, in: Proceedings of the 1st ACM Workshop on Audio and Music Computing Multimedia, ACM, 2006, pp. 21–26.

[71] Anil K. Jain, Data clustering: 50 years beyond k-means, Pattern Recognition Letters 31 (8) (2010) 651–666.

[72] Anil K. Jain, M. Narasimha Murty, Patrick J. Flynn, Data clustering: a review, ACM Computing Surveys (CSUR) 31 (3) (1999) 264–323.

[73] Yihong Gong, Wei Xu, Machine Learning for Multimedia Content Analysis, vol. 30, Springer, 2007.

[74] John A. Hartigan, Manchek A. Wong, Algorithm as 136: A k-means clustering algorithm, Journal of the Royal Statistical Society. Series C (Applied Statistics) 28 (1) (1979) 100–108.

[75] Sanjoy Dasgupta, The Hardness of k-Means Clustering, Department of Computer Science and Engineering, University of California, San Diego, 2008.

[76] Meena Mahajan, Prajakta Nimbhorkar, Kasturi Varadarajan, The planar k-means problem is np-hard, in: WALCOM: Algorithms and Computation, Springer, 2009, pp. 274–285.

[77] Robert Tibshirani, Guenther Walther, Trevor Hastie, Estimating the number of clusters in a data set via the gap statistic, Journal of the Royal Statistical Society: Series B (Statistical Methodology) 63 (2) (2001) 411–423.

[78] L. Vendramin, R. Campello, E.R. Hruschka, On the comparison of relative clustering validity criteria, in: SIAM International Conference on Data Mining, 2009, pp. 733–744.

[79] Takio Kurita, An efficient agglomerative clustering algorithm using a heap, Pattern Recognition 24 (3) (1991) 205–209.

[80] Sabyasachy Choudhury, M. Narasimha Murty, A divisive scheme for constructing minimal spanning trees in coordinate space, Pattern Recognition Letters 11 (6) (1990) 385–389.

[81] Charles T. Zahn, Graph-theoretical methods for detecting and describing gestalt clusters, IEEE Transactions on Computers 100 (1) (1971) 68–86.

[82] Andrew Y. Ng, Michael I. Jordan, Yair Weiss, et al., On spectral clustering: analysis and an algorithm, Advances in Neural Information Processing Systems 2 (2002) 849–856.

[83] Ulrike Von Luxburg, A tutorial on spectral clustering, Statistics and Computing 17 (4) (2007) 395–416.

[84] Stella X. Yu, Jianbo Shi, Multiclass spectral clustering, in: Proceedings of the Ninth IEEE International Conference on Computer Vision, IEEE, 2003, pp. 313–319.

[85] Wei Xu, Xin Liu, Yihong Gong, Document clustering based on non-negative matrix factorization, in: Proceedings of the 26th Annual International ACM SIGIR Conference on Research and Development in Informaion Retrieval, ACM, 2003, pp. 267–273.

[86] D. Seung, L. Lee, Algorithms for non-negative matrix factorization, Advances in Neural Information Processing Systems 13 (2001) 556–562.

[87] Chris Ding, Xiaofeng He, Horst D. Simon, On the equivalence of nonnegative matrix factorization and spectral clustering, in: Proceedings of SIAM Data Mining Conference Number 4, 2005, pp. 606–610.

[88] Farial Shahnaz, Michael W. Berry, V. Paul Pauca, Robert J. Plemmons, Document clustering using nonnegative matrix factorization, Information Processing and Management 42 (2) (2006) 373–386.

[89] Tuomas Virtanen, Monaural sound source separation by nonnegative matrix factorization with temporal continuity and sparseness criteria, IEEE Transactions on Audio, Speech, and Language Processing 15 (3) (2007) 1066–1074.

[90] Glenn W. Milligan, Martha C. Cooper, An examination of procedures for determining the number of clusters in a data set, Psychometrika 50 (2) (1985) 159–179.

[91] Sandrine Dudoit, Jane Fridlyand, A prediction-based resampling method for estimating the number of clusters in a dataset, Genome Biology 3 (7) (2002), research0036.

[92] Peter J. Rousseeuw, Silhouettes: a graphical aid to the interpretation and validation of cluster analysis, Journal of Computational and Applied Mathematics 20 (1987) 53–65.

[93] Olivier Chapelle, Bernhard Schölkopf, Alexander Zien, et al., Semi-Supervised Learning, vol. 2, MIT Press, Cambridge, 2006.

[94] Nizar Grira, Michel Crucianu, Nozha Boujemaa, Unsupervised and semi-supervised clustering: a brief survey, A Review of Machine Learning Techniques for Processing Multimedia Content 1 (2005) 9–16.

[95] Jose Pardo, Xavier Anguera, Chuck Wooters, Speaker diarization for multiple-distant-microphone meetings using several sources of information, IEEE Transactions on Computers 56 (9) (2007) 1212–1224.

[96] Hiroaki Sakoe, Seibi Chiba, Dynamic programming algorithm optimization for spoken word recognition, IEEE Transactions on Acoustics, Speech and Signal Processing 26 (1) (1978) 43–49.

[97] Fumitada Itakura, Minimum prediction residual principle applied to speech recognition, IEEE Transactions on Acoustics, Speech and Signal Processing 23 (1) (1975) 67–72.

[98] Richard Bellman, Dynamic programming and lagrange multipliers, The Bellman Continuum: A Collection of the Works of Richard E. Bellman, World Scientific, 1986, p. 49.

[99] Temple F. Smith, Michael S. Waterman, Identification of common molecular subsequences, Journal of Molecular Biology 147 (1) (1981) 195–197.

[100] Joan Serra, Emilia Gómez, Perfecto Herrera, Xavier Serra, Chroma binary similarity and local alignment applied to cover song identification, IEEE Transactions on Audio, Speech, and Language Processing 16 (6) (2008) 1138–1151.

[101] Pascal Ferraro, Pierre Hanna, Laurent Imbert, Thomas Izard, et al., Accelerating query-by-humming on gpu, in: ISMIR, 2009, pp. 279–284.

[102] Pierre Hanna, Pascal Ferraro, Matthias Robine, On optimizing the editing algorithms for evaluating similarity between monophonic musical sequences, Journal of New Music Research 36 (4) (2007) 267–279.

[103] Mihalis Psarakis, Aggelos Pikrakis, Giannis Dendrinos, Fpga-based acceleration for tracking audio effects in movies, in: 2012 IEEE 20th Annual International Symposium on Field-Programmable Custom Computing Machines, FCCM, IEEE, 2012, pp. 85–92.

[104] Aggelos Pikrakis, Audio thumbnailing in video sharing sites, in: 2012 Proceedings of the 20th European Signal Processing Conference, EUSIPCO, 2012, pp. 1284–1288, EURASIP.

[105] Michael A. Casey, Remco Veltkamp, Masataka Goto, Marc Leman, Christophe Rhodes, Malcolm Slaney, Content-based music information retrieval: current directions and future challenges, Proceedings of the IEEE 96 (4) (2008) 668–696.

[106] Tao Li, Mitsunori Ogihara, Toward intelligent music information retrieval, IEEE Transactions on Multimedia 8 (3) (2006) 564–574.

[107] Tao Li, Mitsunori Ogihara, Qi Li, A comparative study on content-based music genre classification, in: Proceedings of the 26th Annual International ACM SIGIR Conference on Research and Development in Informaion Retrieval, ACM, 2003, pp. 282–289.

[108] Thomas Lidy, Andreas Rauber, Evaluation of feature extractors and psycho-acoustic transformations for music genre classification, in: Proceedings of ISMIR, 2005, pp. 34–41.

[109] Juan José Burred, Alexander Lerch, A hierarchical approach to automatic musical genre classification, in: Proceedings of the Sixth International Conference on Digital Audio Effects 03, 2003.

[110] Avery Wang, The shazam music recognition service, Communications of the ACM 49 (8) (2006) 44–48.

[111] Asif Ghias, Jonathan Logan, David Chamberlin, Brian C. Smith, Query by humming: musical information retrieval in an audio database, in: Proceedings of the Third ACM International Conference on Multimedia, ACM, 1995, pp. 231–236.

[112] Yunyue Zhu, Dennis Shasha, Warping indexes with envelope transforms for query by humming, in: Proceedings of the 2003 ACM SIGMOD International Conference on Management of Data, ACM, 2003, pp. 181–192.

[113] Matthew Cooper, Jonathan Foote, Elias Pampalk, George Tzanetakis, Visualization in audio-based music information retrieval, Computer Music Journal 30 (2) (2006) 42–62.

[114] Elias Pampalk, Andreas Rauber, Dieter Merkl, Content-based organization and visualization of music archives, in: Proceedings of the 10th ACM International Conference on Multimedia, ACM, 2002, pp. 570–579.

[115] Ning-Han Liu, Music selection interface for car audio system using SOM with personal distance function, EURASIP Journal on Audio, Speech, and Music Processing 2013 (1) (2013) 20.

[116] Stefan Leitich, Martin Topf, Globe of music-music library visualization using geosom, in: ISMIR, 2007, pp. 167–170.

[117] Andreas Rauber, Elias Pampalk, Dieter Merkl, The som-enhanced jukebox: organization and visualization of music collections based on perceptual models, Journal of New Music Research 32 (2) (2003) 193–210.

[118] Andreas Rauber, Elias Pampalk, Dieter Merkl, Using psycho-acoustic models and self-organizing maps to create a hierarchical structuring of music by sound similarity, in: Proceedings of ISMIR, 2002, pp. 71–80.

[119] Jiajun Zhu, Lie Lu, Perceptual visualization of a music collection, in: IEEE International Conference on Multimedia and Expo, ICME 2005, IEEE, 2005, pp. 1058–1061.

[120] John C. Platt, Fast embedding of sparse music similarity graphs, Advances in Neural Information Processing Systems 16 (2004) 571578.

[121] Chin-Han Chen, Ming-Fang Weng, Shyh-Kang Jeng, Yung-Yu Chuang, Emotion-based music visualization using photos, in: Advances in Multimedia Modeling, Springer, 2008, pp. 358–368.

[122] Yi-Hsuan Yang, Yu-Ching Lin, Ya-Fan Su, Homer H. Chen, A regression approach to music emotion recognition, IEEE Transactions on Audio, Speech, and Language Processing 16 (2) (2008) 448–457.

[123] Yi Hsuan Yang, Yu Ching Lin, Heng Tze Cheng, Homer H. Chen, Mr. emo: music retrieval in the emotion plane, in: Proceedings of the 16th ACM International Conference on Multimedia, ACM, 2008, pp. 1003–1004.

[124] Luke Barrington, Reid Oda, Gert R.G. Lanckriet, Smarter than genius? Human evaluation of music recommender systems, in: ISMIR, vol. 9, 2009, pp. 357–362.

[125] Hung-Chen Chen, Arbee L.P. Chen, A music recommendation system based on music data grouping and user interests, in: CIKM, vol. 1, 2001, pp. 231–238.

[126] Kazuyoshi Yoshii, Masataka Goto, Kazunori Komatani, Tetsuya Ogata, Hiroshi G. Okuno, Hybrid collaborative and content-based music recommendation using probabilistic model with latent user preferences, in: ISMIR, vol. 6, 2006, p. 7.

[127] Anssi P. Klapuri, Automatic music transcription as we know it today, Journal of New Music Research 33 (3) (2004) 269–282.

[128] Juan Pablo Bello, Giuliano Monti, Mark B. Sandler, et al., Techniques for automatic music transcription, in: ISMIR, 2000.

[129] Roger B. Dannenberg, Ning Hu, Discovering musical structure in audio recordings, in: Music and Artificial Intelligence, Springer, 2002, pp. 43–57.

[130] Marl Levy, Marl Sandler, Structural segmentation of musical audio by constrained clustering, IEEE Transactions on Audio, Speech, and Language Processing 16 (2) (2008) 318–326.

[131] Fabien Gouyon, Anssi Klapuri, Simon Dixon, Miguel Alonso, George Tzanetakis, Christian Uhle, Pedro Cano, An experimental comparison of audio tempo induction algorithms, IEEE Transactions on Audio, Speech, and Language Processing 14 (5) (2006) 1832–1844.

[132] Simon Dixon, Evaluation of the audio beat tracking system beatroot, Journal of New Music Research 36 (1) (2007) 39–50.

[133] Anssi P. Klapuri, Antti J. Eronen, Jaakko T. Astola, Analysis of the meter of acoustic musical signals, IEEE Transactions on Audio, Speech, and Language Processing 14 (1) (2006) 342–355.

[134] Daniel P.W. Ellis, Graham E. Poliner, Identifying cover songs' with chroma features and dynamic programming beat tracking, in: IEEE International Conference on Acoustics, Speech and Signal Processing, ICASSP 2007, vol. 4, IEEE, 2007, pp. IV–1429.

[135] Wei Chai, Structural analysis of musical signals via pattern matching, in: Proceedings of the 2003 IEEE International Conference on Acoustics, Speech, and Signal Processing, ICASSP'03, vol. 5, IEEE, 2003, pp. V–549.

[136] Jouni Paulus, Anssi Klapuri, Music structure analysis using a probabilistic fitness measure and a greedy search algorithm, IEEE Transactions on Audio, Speech, and Language Processing 17 (6) (2009) 1159–1170.

[137] Jonathan Foote, Visualizing music and audio using self-similarity, in: Proceedings of the Seventh ACM International Conference on Multimedia, Part 1, ACM, 1999, pp. 77–80.

[138] Aggelos Pikrakis, Iasonas Antonopoulos, Sergios Theodoridis, Music meter and tempo tracking from raw polyphonic audio, in: Proceedings of the International Conference on Music Information Retrieval, ISMIR, 2004.

[139] Douglas Eck, Thierry Bertin-Mahieux, Paul Lamere, Autotagging music using supervised machine learning, in: ISMIR, 2007, pp. 367–368.

[140] Ella Bingham, Heikki Mannila, Random projection in dimensionality reduction: applications to image and text data, in: Proceedings of the Seventh ACM SIGKDD International Conference on Knowledge Discovery and Data Mining, ACM, 2001, pp. 245–250.

[141] Svante Wold, Kim Esbensen, Paul Geladi, Principal component analysis, Chemometrics and Intelligent Laboratory Systems 2 (1) (1987) 37–52.

[142] Ian Jolliffe, Principal Component Analysis, Wiley Online Library, 2005.

[143] Keinosuke Fukunaga, Introduction to Statistical Pattern Recognition, Elsevier, 1990.

[144] Teuvo Kohonen, Self-Organizing Maps, vol. 30, Springer, 2001.

[145] Joshua B. Tenenbaum, Vin De Silva, John C. Langford, A global geometric framework for nonlinear dimensionality reduction, Science 290 (5500) (2000) 2319–2323.

[146] Chih-Chung Chang, Chih-Jen Lin, LIBSVM: a library for support vector machines, ACM Transactions on Intelligent Systems and Technology 2 (2011) 27:1–27:27. <http://www.csie.ntu.edu.tw/~cjlin/libsvm>.

[147] George Tzanetakis, Perry Cook, Marsyas: a framework for audio analysis, Organised Sound 4 (3) (2000) 169–175.

[148] George Tzanetakis, Marsyas-0.2: a case study in implementing music information retrieval systems, Intelligent Music Information Systems. IGI Global (2007).

[149] Davis E. King, Dlib-ml: a machine learning toolkit, Journal of Machine Learning Research 10 (2009) 1755–1758.

[150] Thierry Bertin-Mahieux, Daniel P.W. Ellis, Brian Whitman, Paul Lamere, The million song dataset, in: Proceedings of the 12th International Conference on Music Information Retrieval, ISMIR 2011, 2011.

INDEX

Printed and bound by CPI Group (UK) Ltd, Croydon, CR0 4YY

03/10/2024

01040414-0015